AF213379

Double Photoionisation Spectra of Molecules

John H. D. Eland and Raimund Feifel

OXFORD
UNIVERSITY PRESS

OXFORD
UNIVERSITY PRESS

Great Clarendon Street, Oxford, OX2 6DP,
United Kingdom

Oxford University Press is a department of the University of Oxford.
It furthers the University's objective of excellence in research, scholarship,
and education by publishing worldwide. Oxford is a registered trade mark of
Oxford University Press in the UK and in certain other countries

© John H. D. Eland and Raimund Feifel 2018

The moral rights of the authors have been asserted

First Edition published in 2018

All rights reserved. No part of this publication may be reproduced, stored in a retrieval system, or transmitted, in any form or by any means, without the prior permission in writing of Oxford University Press, or as expressly permitted by law, by licence or under terms agreed with the appropriate reprographics rights organization. Enquiries concerning reproduction outside the scope of the above should be sent to the Rights Department, Oxford University Press, at the address above

You must not circulate this work in any other form
and you must impose this same condition on any acquirer

Published in the United States of America by Oxford University Press
198 Madison Avenue, New York, NY 10016, United States of America

British Library Cataloguing in Publication Data
Data available

Library of Congress Control Number: 2017943623

ISBN 978–0–19–878898–0

Links to third party websites are provided by Oxford in good faith and
for information only. Oxford disclaims any responsibility for the materials
contained in any third party website referenced in this work.

Contents

Preface

The primary purpose of this volume is to provide the community of physical chemists and molecular physicists with a new resource on the resolved electronic spectra of doubly charged molecules. The spectra have been made accessible by the recent development of new techniques for multiple electron analysis and detection, together with pulsed extreme ultraviolet and soft X-ray light sources. Both the authors have been intimately involved in these developments in the laboratory and at electron storage rings, and now propose to make the techniques and results widely available to students and researchers by means of the present work. To this end, the main body of the book is a compilation of more than 70 double photoionisation spectra of compounds from diatomic through to small polyatomic molecules. The Introduction, which explains the concepts and techniques, is written at a level that is intended to make the subject accessible to both physics and chemistry students embarking on research in the area of molecular spectroscopy. The spectra provide a basis for advanced experiments and calculations on the ions, particularly on the dynamics of their dissociations and reactions with other species. They also illustrate how the molecular orbital model of electronic structure is tested and stretched by the phenomena of double ionisation. The collection contains double ionisation spectra of all the major constituents of planetary atmospheres and prototypes of several classes of organic molecules, some involved in biological processes. Double ionisation energies of many molecules are given here in a compact and convenient form. Acquisition of ionisation spectra implies the location in energy of onsets and peaks, which normally equate to ionisation energies, but there is some ambiguity over their exact meaning in double ionisation, particularly when no stable molecular doubly charged ions exist. These complications are carefully examined. In addition to conventional double ionisation where two electrons are removed from molecular valence shells, the text includes the more recently opened fields of core–core and core–valence double ionisation of molecules. Appendices cover the mechanisms of double photoionisation in detail, practical data for spectrum calibration and a brief survey of the major theoretical methods used to calculate ionisation energies and spectra. The presentation is non-mathematical and the background necessary to make use of the book is no more than a second year university course in chemistry, molecular physics or another science, providing an introduction to the concepts of molecular electronic structure.

The spectra presented here were gathered mainly during the years since 2003, and while many have appeared in learned journals, some have not been published before and are shown here for the first time. All the spectra have been newly prepared from original data. In the interests of brevity in a field where large teams of scientists are usually involved, the author lists given in the references are quoted in full only when there are three authors or fewer. We apologise to any authors who feel unhappy at the omission of their names. Other omissions and errors in the text are entirely ours, despite the efforts of the people who have helped us.

Acknowledgements

Many people have been involved in creation of the material presented in this book, including graduate students at Oxford, Uppsala, Stockholm, and Gothenburg, and we gratefully acknowledge their contribution. We particularly thank Leif Karlsson for reading earlier forms of the manuscript and making many useful suggestions to improve it. Very special thanks are due to Ieva Eland for meticulous and repeated reading of all the chapters and correcting many errors and infelicities of orthography and expression.

Archive of digital spectra

All the double ionisation spectra in this book, including core–valence spectra, are available in digital form at www.oup.com/doublephotoionisationspectra2018 with the following Username and Password:

Username: OUPBooks
Password: Eland&Feifel

The individual spectra in the archive are named with the following conventions.

Double ionisation spectra have the conventional chemical name of the compound followed by capital E (for energy), the photon energy in eV to the nearest integer and termination .PLO (for plot). In the book itself, the number of points is given first followed by listing of the ionisation energy (eV) and the number of counts at each point. The exact photon energy and other experimental conditions can be found with the relevant figure in the book. For instance, the double ionisation spectrum of bromine at 40.81 eV photon energy is listed as Br2E41.PLO and details are in Fig. 2.7.1. For larger molecules, the chemical name is written in full, for example FerroceneE27.PLO for the spectrum of ferrocene at 26.91 eV photon energy, as in Fig. 5.9.1.

Core–valence spectra are listed with a similar naming convention. After the compound's name comes CV (core–valence) followed by the chemical symbol for the atom on which the core hole is located. For carbon dioxide, for example, when the core hole is in the C1s orbital, the name is CO2CVC.PLO and if the hole is in O1s, the name is CO2CVO.PLO. The files are in the form of the number of points followed by a listing of the energy and number of counts at each point, but here the energy in eV is measured from the energy of the simple core hole, as given in the relevant figure or its caption.

Introduction

Introduction

Molecular spectroscopy has traditionally concentrated on neutral molecules, the dominant constituents of our world. But neutral molecules can gain or lose electrons to form molecular ions, and these rarer species are absolutely vital as intermediates in chemical reactions, in outer planetary atmospheres, in astrophysical environments such as stellar outflows and interstellar clouds and in the whole realm of the 'fourth state of matter', plasmas. In the universe as a whole, positive ions are overwhelmingly more abundant than neutral species, though we should admit that most of these are atomic rather than molecular. This book is concerned with positive molecular ions which carry two or more positive charges, whose spectra have become accessible only very recently. Spectra of molecular ions with just one positive charge, also called monocations, were also relatively unknown until about 50 years ago, because the only methods available to measure them were indirect and limited to just a few of their electronic states. With the advent of photoelectron spectroscopy in the 1960s that situation was completely changed, as the new technique allowed complete spectra of valence electron ionisations for each molecule to be revealed in a single measurement. There are now compendia of the spectra of many monocations available, and these show both the ground state and the excited states of each ion, in contrast to conventional electronic spectra of neutral molecules which show only excited states. In the early years of the twenty-first century a new technique was invented, which does the same for doubly positive ions, also called dications, as photoelectron spectroscopy did for monocations, and which is the main topic of this book. It could be called 'double photoelectron spectroscopy', but as we shall see, the name in common use is an acronym, more directly related to the techniques employed than to the outcome of the measurements. This book is a compendium of the spectra of small and intermediate-sized dications, the doubly charged species most likely to be encountered in important physical environments. The many examples and the general principles involved in the analysis are intended to provide a starting point for new techniques that go further in investigating the properties of the dications and the physics of their formation and destruction. For some of the molecules, calculations of their electronic structure have been performed by several groups, and references but no details are given. Where there are no useful calculations we include some empirical estimates with discussion and analysis based on the molecular orbital structure, as revealed in conventional photoelectron spectra. Descriptions of current theoretical methods and of the results to be expected from them are included in Appendix 3.

Double Photoionisation Spectra of Molecules. John H. D. Eland and Raimund Feifel.
© John H. D. Eland and Raimund Feifel, 2018. Published in 2018 by Oxford University Press.
DOI 10.1093/oso/9780198788980.001.0001

1 Basis of the TOF-PEPECO technique

In photoelectron spectroscopy, molecules in the gas phase interact with extreme ultra-violet light of a single energy $h\nu$, ejecting electrons with a definite kinetic energy (KE), which can be measured. The classical Einstein equation for the photoelectric effect can be written in terms of the ionisation energy (IE) as

$$\mathrm{IE} = h\nu - \mathrm{KE}$$

where IE can also be understood as the binding energy of a particular electron in an atom or molecule. The spectrum of electron energies, that is, the intensity of photoelectrons plotted against their energy, is a photoelectron spectrum and shows all the electronic and vibrational states of the positive ion accessed at that photon energy. To a surprisingly good approximation, the ionisation energy to each ion state represents the orbital binding energy of the electron that has been ejected. The idea behind this brilliant technique seems to have come to several workers in different countries independently (F.I. Vilesov in Russia, W.C. Price and D.W. Turner in England and K. Siegbahn in Sweden). But the breakthroughs that made the technique practical were the development of ionising light sources, the helium lamp in London and X-ray sources in Sweden. The helium lamp, developed particularly by Turner in London, provides photons of 21.22 eV photon energy, which is sufficient to reveal all the valence electron ionisations of most molecules. The energy spread within the main helium atomic line (HeIα) is narrow enough to provide at least vibration level resolution. The Al and Mg Kα X-ray lines used by Siegbahn also reveal the inner valence electrons and the core electrons, but with relatively poor energy resolution because of the width of the X-ray lines. To clarify the meaning here and elsewhere, the term *valence* or *outer valence* refers loosely to the electrons potentially involved in chemical bonding, particularly electrons in the outermost orbitals such as those made from 2p atomic orbitals of first row atoms. The term *inner valence* then refers to orbitals constructed from deeper atomic orbitals such as the 2s atomic orbitals in first row atoms. All deeper orbitals such as those based on 1s atomic orbitals are called *core* molecular orbitals.

Two major difficulties arise in an attempt to apply the techniques of photoelectron spectroscopy to double photoionisation. Firstly, the probability of a single photon ejecting two rather than just one electron is very small, though not zero. In the familiar orbital (one-electron) model, double ionisation is completely forbidden as a single-step process if the orbitals themselves do not change when an electron is removed. Neither this 'frozen orbital' model nor the one-electron model of orbitals is strictly valid, and it is the breakdown of the simple models that makes double photoionisation by single photons possible at all. The second difficulty is that because two electrons are ejected in the process

$$\mathrm{m} + h\nu \rightarrow \mathrm{m}^{++} + \mathrm{e}^- + \mathrm{e}^-$$

the Einstein equation must be modified to read

$$\mathrm{DIE} = h\nu - (\mathrm{KE}_1 + \mathrm{KE}_2)$$

where DIE stands for double ionisation energy and KE_1 and KE_2 are measured energies of the two electrons. The very small momentum, and therefore energy given to the ion by recoil, can almost always be neglected in practice and it is neglected here. For the equation to be practically useful, the two electrons whose kinetic energy is measured must both come from the same ionisation event, as only their energy sum is well defined. In each double ionisation event where a single photon interacts with a single molecule the two electrons can share the available energy between them in an arbitrary manner. To identify such pairs of electrons born together in single ionisation events, we must select electrons detected, as they have been born, at almost the same time. Electrons produced by single ionisation in different events have only random times of birth and detection, so true pairs stand out because of their correlation in time. Such pairs with a definite time correlation are said to be in *coincidence* with each other. In actual measurement, we must measure the times of detection of all electrons; for two coincident electrons, born together at the same instant, the relative arrival times at one or more detectors, determined by their velocities and the path length that each must follow, are always time-correlated.

The earliest attempts at photoelectron–photoelectron coincidence measurements used two conventional electron energy analysers, each of which registered a single energy at a time, viewing a common ionisation region. Measurements were extremely slow even at low resolution, so this method proved impractical. Better success was had by concentrating on zero-energy (also called 'threshold') electrons, which can be detected with very high efficiency, using synchrotron radiation from a storage ring as a tuneable far-ultraviolet light source. This technique produces good spectra of doubly charged ions, but the measurements require lengthy runs to achieve good statistics, and runtime at storage rings is very limited.

Three breakthroughs which allowed spectra of doubly charged ions to be measured routinely were the adaptation of a device called a magnetic bottle for use as a time-of-flight (TOF) electron spectrometer [1], the development of a source of short pulses of high energy (extreme ultraviolet or soft X-ray) light [2] and the evolution of electronic devices to record multiple signals very rapidly. These three developments when put together gave birth to the technique now called *magnetic bottle time-of-flight photoelectron-photoelectron coincidence spectroscopy*, or TOF-PEPECO for short. Its principle is that ionisation of target molecules is caused by a short pulse of light at one end of the bottle and the ejected electrons are guided to a detector at the other end, where their arrival times relative to the light pulse are recorded. From the time between the light pulse and the arrival of an electron at the detector the electron velocity is derived and hence the electron energy. In double photoionisation two electrons are emitted at the same time, and both arrive at the detector within a very short interval, a few microseconds. So long as the rate of single ionisation, determined by the gas pressure and light intensity, is low enough that two separate molecules are very rarely ionised on a single light pulse, the detection of two electrons after one pulse is sufficient to recognise a true coincidence. As both electron energies are known from their arrival times, the double ionisation energy can be derived according to the modified Einstein equation. Another light source, synchrotron radiation from electron storage rings, was brought into service soon after the initial developments and is now a major aspect of the technique. Details of these devices and techniques are given in Section 6.

2 Other techniques for spectra of doubly charged ions

There were several ways to measure spectra of doubly charged ions before the advent of TOF-PEPECO. In *Auger electron spectroscopy* [3] a high-energy electron or X-ray photon ejects an inner shell electron from the molecule, for instance a 1s electron from a carbon atom. An outer electron then falls into the inner hole and the energy so liberated ejects a second electron from the same or another outer orbital. The result is a doubly charged ion and an Auger electron whose energy, when subtracted from the energy of the initial hole state, gives the energy of a state of the ion. One problem is that inner shell hole states have short lifetimes τ and are therefore energetically broad, with width ΔE according to the uncertainty relationship:

$$\tau\Delta E \approx h/2\pi.$$

Every line in an Auger electron spectrum has this same width which is always comparable to or larger than vibrational level spacings, so the Auger spectra have limited resolution. Many Auger spectra are also complicated by overlapping features because it is often impossible to create just a single inner shell hole state; each hole contributes its own Auger spectrum. Nevertheless, Auger spectra of many atoms and molecules have been measured over the years and provide a very valuable resource. References to such spectra are given in the main body of the text wherever they are available for the molecules we cover. The resolution problem in Auger spectroscopy can be overcome if the energies of both the photoelectron from creation of the initial hole and the Auger electron from each ionisation event are measured. This requires a coincidence measurement and it has only recently become possible, mainly by the technique described here or by a closely related one.

In *double charge transfer (DCT) spectroscopy* [4] a positively charged atomic ion of well-defined velocity hits a neutral molecule and abstracts two electrons from it, leaving a doubly charged ion and itself becoming a negative ion. The angle by which the outgoing negative ion is deflected in the collision while still being detected is tightly constrained to near zero, and the ion's energy spectrum is measured. The kinetic energy lost by the projectile in the charge-transfer collision is the energy needed to create a doubly charged molecule in its various accessible states, and so constitutes the desired spectrum. Because the probability of double charge transfer in the collision is very small, this is a difficult and time-consuming experiment, but it has been developed to a high degree and applied in numerous cases. The spectra produced have the advantage that by choice of projectile, the experimenter can control the spin (multiplicity) of the molecular ions created. They have the disadvantage that the transition probability is a strongly diminishing function of the energy transfer, so higher-energy states of the product doubly charged ions are not well seen. Again, references to useful spectra taken by this method are given later.

The technique with acronym *TPEsCO (threshold photoelectrons coincidence)* [5] is closely related to the TOF-PEPECO technique. The phrase *threshold electrons* means electrons of near-zero kinetic energy; several ingenious techniques have been devised to detect such electrons with very high efficiency. In TPEsCO, pairs of such threshold electrons are recorded as the energy of ionising light is scanned. Since the electrons have zero energy, the intensity of the pair signal at each photon energy $h\nu$ is the spectrum of doubly charged ions. This technique can have excellent resolution, but because the cross-section for double ionisation by emission of zero-energy electrons is very small, and the spectrum must be measured point by point, it is very time consuming. This is a serious disadvantage, as the only practical source of tuneable light in the necessary EUV (extreme ultraviolet) and X-ray region is synchrotron radiation from a storage ring, and access to such sources is severely limited.

The last technique to be mentioned here is the measurement of kinetic energy distributions in dissociation of doubly charged ions from diatomic molecules. When a single molecule AB is turned into a dication AB^{2+} by an energetic collision process, it is initially in a defined quantum state with a definite energy. Some such states are stable on the timescale of laboratory experiments, but the majority dissociate rapidly into a pair of atomic cations $A^+ + B^+$. The kinetic energies of those two cations can be measured with very good resolution, and added together. Because the ions are atomic and have few accessible states, the energy of the initial dication can be inferred by adding the total kinetic energy to the known electronic energies of the products. Although this is an indirect and quite arduous procedure which can be applied to rather few molecules, it has yielded exquisite spectra of predissociating states of a few diatomic species [6] and so merits honourable mention.

3 Ionisation energies

An important parameter that comes from these measurements is the energy required to create a doubly charged ion from the neutral molecule in its ground state. This is variously called the double ionisation potential, the double ionisation energy (DIE) or the appearance energy of the doubly charged ion. Whatever the name, care is needed in applying it to experimental results from electron spectra, especially in the many cases where the doubly charged ions have no ‘stable’ or metastable states with lifetimes longer than a vibration period. As a starting point, we define the *vertical double ionisation energy* as the energy of the most probable (intense) transition in which the atoms remain in their equilibrium positions when becoming doubly ionised. Such transitions are called vertical because they are represented by vertical lines between curves for different states in a potential energy diagram. They can be measured by TOF-PEPECO when the double ionisation is direct, that is, it happens in a single step. Even if the lowest dication state is not a bound molecular ion with a well in its potential energy curve, but rather purely repulsive, this definition can still be used. Only where vibrationally resolved features are seen can *adiabatic double ionisation energies* be determined, defined as the energies of transitions from the lowest vibrational level ($v''=0$) in the neutral ground state to $v'=0$ in each electronic state of the dication. Many molecules have both bound and unbound dication states accessible to double photoionisation.

A complication arises from the frequent existence of *indirect* pathways by which two electrons can be ejected, and these are discussed later in this chapter, and in more detail in Appendix 1.

4 Orbital configurations and dication states

In single ionisation of the outer electrons (valence electrons) of a molecule there is a simple relationship between the photoelectron spectrum and the energies of the occupied molecular orbitals, expressed by the approximation known as Koopmans' theorem. For each doubly occupied orbital of a closed-shell molecule we see one band in the spectrum with ionisation energy roughly equal to the orbital binding energy. In double ionisation the situation is more complicated, but the orbital picture is still useful. If two electrons are taken from the same doubly occupied orbital, we get a singlet state, but one each taken from two different orbitals gives both a singlet and a triplet. Two electrons taken from a full π-orbital in a symmetrical linear molecule give three states according to the Pauli principle, $^3\Sigma_g^-$, $^1\Delta_g$ and $^1\Sigma_g^+$. So from knowledge of the orbitals in a target molecule we can work out how many simple dication states should exist, and make a guess at their energies by summing the ionisation energies of the orbitals involved and adding a Coulomb repulsion term e^2/r_{ij} dependent on the size of the molecule and reasonable assumptions about where the charges are located. For nitrogen, N_2, for instance, the ionisation energy from the outermost orbital $2p\sigma_g$ is 15.6 eV, the bond length is 1.1 Å and the crudely predicted double ionisation energy to form a $^1\Sigma_g^+$ state is 44.3 eV. The observed energy is 42.9 eV. Such calculations, which can be refined, serve as a rough guide to identification of structures in the spectra with the expected electronic states. There is no restriction in photoionisation on the symmetries of the states to be formed, so all combinations of orbitals and spins involving changes to just two electrons are possible. Their relative probabilities and intensities in the spectra are much more difficult to predict and are at present to be determined by experiment.

In ordinary photoelectron spectra showing single ionisation the majority of states which appear as strong bands correspond to removing one electron from an occupied orbital and are called one-hole (1h) states for short. For some molecules, weaker bands appear in the valence region, from ionisation of one electron with simultaneous excitation of another; the states involved are called two-hole one-particle (2h-1p) states. Such states are common in the inner valence region for most molecules, and if their energies are high enough they can be important intermediates in double ionisation. In double photoionisation spectra similarly, the strongest bands come from simple two-hole states, and other bands are attributed to three-hole one-particle (3h-1p) states. These 3h-1p states, for which we shall use the abbreviations from now on, can make a substantial contribution to double photoionisation spectra. The extent of their contribution can be judged at present only by theoretical methods, in particular by Green's function calculations (see Appendix 3).

5 Ionisation pathways and their effects

From examination of the distributions of electron energies in double photoionisation we know that much of the creation of final doubly charged ions goes by *indirect* rather than *direct* pathways. In the direct process, photoionisation happens in a single step and the two outgoing electrons share the available excess energy in a statistical fashion; each one can have any energy from zero up to the maximum available, E_{max}, which is the photon energy minus the ionisation energy. If the first electron has energy E_1, the second electron must have the complementary energy $E_{max} - E_1$ and each has a smooth intensity distribution as a function of energy. In the indirect process, by contrast, a distinct but short-lived state of the singly charged ion is formed first, which later emits a second electron leaving the dication. The normal Auger process, where the first photoelectron is ejected from an inner shell, is the prime example of indirect double ionisation. The individual electron spectra are not smooth distributions, but have discreet peaks at characteristic but complementary energies. The energies are complementary because their sum must be equal to the difference between the photon energy and the energy of a definite doubly charged state. Such indirect processes can complicate the determination of double ionisation energies and also affect relative state intensities in the final spectrum. If the molecular geometry of the intermediate monocation state, in bond lengths and bond angles, is the same as that of the neutral ground state there will be no effect on the dication spectrum, but if the structure is different or the state is unbound, there can be significant effects.

The energy of charge-separated products, formally $A^+ + B^+$, is always lower than the energy of the molecular dication AB^{2+}, so if an intermediate monocation state has begun to dissociate to $A^+ + B^*$ the final products $A^+ + B^+$ can be formed at a lower ionisation energy than the molecular AB^{2+}. Such stepwise processes are very common and occur even in small strongly bound molecules such as CO. As photon energy rises, the probability of direct double photoionisation increases from near zero at threshold over a range of a few hundred eV, whereas the probability of the indirect pathways through single photoionisation remains constant or falls. Thus the proportion of indirect double ionisation falls as photon energy increases, and its effects in the spectra may be recognised by their energy dependence. An empirical observation from many TOF-PEPECO electron distributions is that the intermediate states involved in indirect pathways to double ionisation are mostly found at energies not far above the double ionisation threshold. This means that they are largely absent in spectra generated exclusively from electron pairs, both of which have energy above an appropriately chosen minimum. Such selective analysis has been applied in deriving a few of the spectra presented here. Restriction to electrons with more than a low minimum energy of 0.5–1 eV is applied rather more frequently, as it also prevents distortion of the spectra by pile-up of background electrons from wall collisions and other unwanted processes at near-zero energy.

Intermediate singly charged ions may not only start to dissociate to singly charged products but may actually succeed in dissociating. The excess internal energy held by the intermediate can be used to break bonds rather than to ionise further, so although the energy transferred from photon to molecular ion is above the double ionisation energy, there may be no double ionisation. The lifetimes τ of such intermediate states can be relatively long, at least equal to a vibration period. This means that their width in energy ΔE will be small, of the order of a tenth of an eV, according to the uncertainty relationship mentioned before,

$$\tau \Delta E \approx h/2\pi,$$

so where they do ionise further they can appear as sharp structures above the minimum double ionisation energy. Even some intermediate states which do ionise further as molecules, or as fragments of molecules, can also have long enough lifetimes to show up as distinct structures in single ionisation spectra above the double ionisation threshold. In the same way, some doubly charged states may be seen in our spectra above the triple ionisation energy, because dissociation has taken precedence over further ionisation or the intermediate lifetime is otherwise sufficiently long. More details of the different possible ionisation pathways are given in Appendix 1.

6 How TOF-PEPECO experiments are done

6.1 Light sources

Incoherent sources To be useful for TOF-PEPECO, a light source must produce radiation with at least enough energy per photon to doubly ionise molecules, which requires a minimum of about 25 eV to see the lowest energy dication state, and 40 eV or more to cover all orbitals in the valence shell. To allow electron timing, the light must be in pulses of no more than a few to a few tens of nanoseconds (ns, 10^{-9} s), which is the major difficulty. Development of a suitable laboratory light source was a key component in the early development of the technique. All the first TOF-PEPECO spectra were taken using lamps producing the atomic emission lines of helium or neon, most commonly the HeII lines at 30.4 and 25.6 nm wavelength (40.81 and 48.37 eV photon energy). The lamps were and still are based on a capillary discharge in the rare gas, driven by the energy stored in a small capacitor charged to a voltage of the order of 10 kV, mounted coaxially to minimise inductance. In these lamps the charge on the capacitor is built up until the voltage becomes high enough to initiate spontaneous breakdown in the capillary, which produces the light flash and discharges the condenser. Positive high voltage is applied to one side of the capacitor to charge it through a resistor until a fast switching device connects the positive end of the capacitor to ground. This creates a very rapidly rising negative voltage at the other end, which is the electrode mounted on the capillary. The gas in the capillary discharge tube breaks down spontaneously when the voltage is high enough and discharges the capacitor; the cycle is repeated at a controlled repetition rate. The simplest external switch is a spark gap in air, but more control is possible at the expense of technical complication with an electronic switch, as illustrated in Fig. 1. A suitable fast high-voltage switch can be made using a hydrogen thyratron or a solid-state circuit. With such devices, pulses of a few ns length can be produced at a repetition rate of the order of 10 kHz. The most important photon energies produced are 40.81 and 48.37 eV from He and 26.81 and 32.69 eV from Ne. A grating monochromator to isolate the different atomic lines is almost indispensable but unless close multiplet lines, such as those from NeII are to be separated, high resolution is not necessary.

A more flexible but less accessible source of pulsed high-energy light is synchrotron radiation from an electron storage ring. In such a source, pulses of electrons are accelerated to very nearly the speed of light (energies >100 MeV) and are constrained

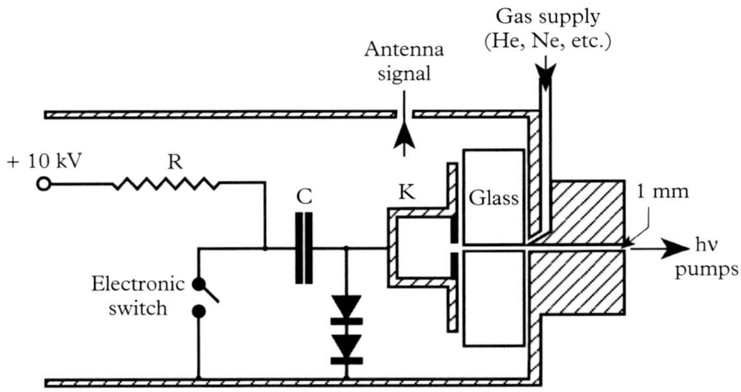

Figure 1 *Schematic diagram of a pulsed atomic discharge lamp for TOF-PEPECO. The condenser C of about 100 pF is charged through the diode chain and resistance R, chosen to match the charging time to the desired repetition rate. Discharge is initiated by closing the electronic switch, which causes a rapidly rising negative potential to appear on the cathode K. When the breakdown voltage is reached, a high-current pulse passes through gas in the glass capillary discharge tube and the antenna picks up a strong positive-going signal for use as a time reference.*

by magnetic fields to circulate in a ring. Wherever the electron path is bent by a magnetic field, radiation with a broad spectrum reaching into the X-ray region is emitted, mainly in the forward direction tangential to the path. These marvellous sources of radiation from the far infrared (THz) up to hard X-rays are generally large national facilities, where individual researchers must apply for runtime, and generally are accorded a few weeks each year. A serious difficulty is that in normal operation the storage rings produce very short pulses of light (sub-nanosecond) but at much too high a repetition rate to be useful for TOF spectroscopy. The time between individual pulses is of the order of 1 ns, whereas electron flight times in even the shortest TOF instruments are about 1,000 times longer. Most rings provide periods of 'single bunch mode' where only a single bunch of electrons circulates in the ring, and here the period between light pulses is of the order of one microsecond; these are the only periods suitable for TOF photoelectron spectroscopy at present. Newer operating regimes including 'few bunch mode' and 'hybrid modes' are being developed and should be suitable for TOF spectroscopy in the future. The higher-energy double photoionisation spectra reported in this book have all been recorded at storage rings operated in single bunch mode. They are sometimes called 'complete' because they cover the whole realm of double ionisation, from its threshold right up to and beyond the triple ionisation energy. The modern synchrotron radiation sources all include what are called 'insertion devices', usually 'undulators', that make the radiation much more intense for a chosen resolution. They are explained further in the next section.

High harmonic generation and free-electron lasers The most powerful light sources in all regions of the spectrum are lasers. Table-top laboratory lasers produce intense collimated beams of light from the near-infrared to the ultraviolet end of the spectrum, but none of them are directly useful for double ionisation spectroscopy in a simple way. To convert their output into the range of soft X-rays and extreme ultraviolet (EUV), where double ionisation can be brought about by single photons, we must use non-linear processes which generate high harmonics. For instance, the twenty-first harmonic of the very common 800 nm infrared laser is at 32.5 eV, a very useful energy. But this is much easier said than done. The conversion process requires very high instantaneous power in the initial laser beam to overcome the inefficiency of the non-linear process of harmonic generation, and this means that very short laser pulses must be used with widths of some tens of femtoseconds (fs, 10^{-15} s). The shortness of the pulses spreads them in energy by the uncertainty principle, and the spread is amplified by the harmonic number. If the harmonic light is used as it is, the energy resolution is compromised and if it is filtered by a monochromator to improve the resolution, the intensity becomes impractically low. For these reasons, little or no spectroscopic work on doubly charged ions has been done using high harmonic generation (HHG) up to now (2017). While HHG is currently limited in spectral energy resolution, it can offer exceptional resolution in timing. Subfemtosecond visualisation of the double ionisation of atomic xenon has been demonstrated [7] using lasers and HHG. But there is another sort of laser that is extremely intense, sufficiently monochromatic and operates in the EUV and X-ray regions, the free-electron laser (FEL). Such sources are based on refined forms of the insertion devices mentioned above, which can be located in an electron storage ring or in a linear accelerator. In a long undulator there are many magnets with alternating field directions which bend the electron beam into a roughly sinusoidal path. For certain wavelengths, the synchrotron light emitted from one crest can be in phase with that emitted from the next crest all along the chain; such light gets amplified coherently and can emerge as a laser beam. Because of relativistic contractions in both time and length in the transformation from the electron's frame of reference to the laboratory frame, the coherent emission is in the EUV or X-ray region, even though the magnet spacing is of the order of centimetres. Of course, there is much more to the operation of undulators and free-electron lasers [8, 9], but that lies beyond the scope of this book. Intense laboratory lasers in the visible or infrared can cause double ionisation by multiphoton processes. The processes are interesting and widely studied, but have not so far contributed to the spectroscopy of doubly charged ions.

6.2 The magnetic bottle TOF spectrometer

The name *magnetic bottle* comes from the fact that a divergent magnetic field can effectively repel electrons coming towards it, so two such fields facing one another in line can confine the particles to a restricted region between them. While magnetic bottles had been used in physics to confine charged particles long before, their use as electron spectrometers was initiated by Kruit and Read [1] only in 1983. The magnetic bottle they developed for use as an electron spectrometer has a strong (ca 1 T or 10 kGauss) divergent magnetic field in the region where ionisation takes place, leading on to a low (ca 10^{-3} T or 10 Gauss) uniform magnetic field inside a solenoid, which guides electrons to a 50 cm distant detector. Photoelectrons are created by a short light pulse and their flight time to the detector is recorded, to be converted later to the electron energy. The special characteristic of this sort of device is that if the fields are configured correctly, more than 90% of all electrons coming from a point source can be sent to a distant detector. This gives the magnetic bottle TOF electron spectrometer an outstanding collection and detection efficiency of at least 50%, meaning that at least half of all electrons created in the source are detected and registered. This is to be compared with a few percent at best for conventional spectrometers. Furthermore, electrons of all energies can be recorded all the time, as there is no slit and no need for scanning through the energies. The first research using such apparatus was spectroscopy of single ionisation, brought about by one or a few photons from pulsed lasers. An advance which helped to inspire the creation of the TOF-PEPECO technique was the development in 1995 of a long, ca 4 m, magnetic bottle TOF spectrometer to record single photoelectrons from laser photodetachment of electrons from negative ions [10]. With such a long magnetic bottle, useful electron energy resolution could be achieved even with 100 ns ionising light pulses, the sort already known to be easily made in the laboratory. The realisation that its high collection efficiency, coupled with the fact that all energies are measured all the time ('multiplex advantage'), would allow electron–electron coincidence measurements to be done far faster and more effectively using this apparatus than by any other method led to the first TOF-PEPECO measurements in 2003 [2]. In the event, pulses less than 5 ns long were generated by the new lamps and the other advantages of the magnetic bottle technique also came to fruition. As a result, most spectra in this collection have been measured using either a 2 m or a 5.6 m magnetic bottle TOF-PEPECO spectrometer. A schematic diagram of the apparatus is given in Fig. 2 and a photograph of the apparatus in use at a synchrotron radiation source is shown as Fig. 3. Almost all the

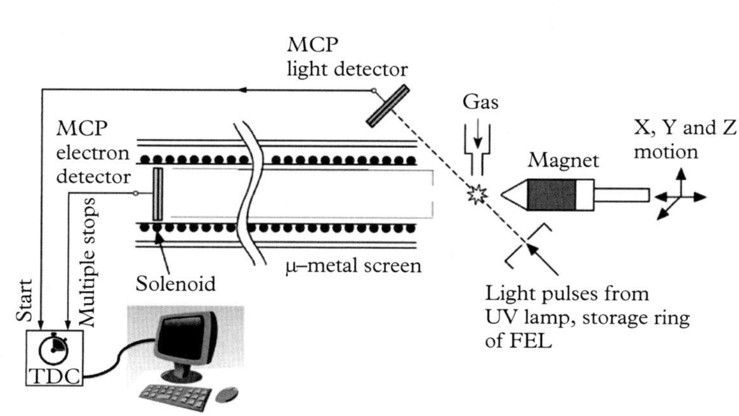

Figure 2 *Schematic diagram of a TOF-PEPECO spectrometer, in roughly the same orientation as that shown in Fig. 3.*

Figure 3 *The 2 m TOF-PEPECO apparatus at work in the experimental hall of the synchrotron radiation facility BESSY 2 in Berlin. The long tube in the centre contains the electron flight path, inside a μ-metal (high-permeability) screen to shield it from the geomagnetic field. At the left end of the tube is the electron detector, followed by the electronics rack and computers. Synchrotron light enters at the same level through a tube at the extreme right of the picture. Below the bench are vacuum pumps and their controllers.*

valence electron double ionisation spectra have been recorded in the laboratory using the 5.6 m instrument, which gives the best electron energy resolution. All the spectra taken at higher photon energies have used the 2 m machine with synchrotron radiation.

Electron detection, timing electronics and calibration In TOF-PEPECO, electrons are detected by an assembly of two or three microchannel plate (MCP) electron multipliers mounted in series; such a detector offers a large sensitive area, subnanosecond timing accuracy and detection efficiency of 50% or better. For each detected electron, the output from the MCP assembly is a pulse of about 10^8 electrons with a duration of a nanosecond or so. These pulses are amplified and fed into a fast time-to-digital converter capable of recording multiple pulse arrivals after a single start signal, or a flash digitiser which records a complete waveform. In the infancy of our technique such devices were scarce and sometimes home-built, but as they are now readily available commercially the construction details are no longer relevant. As the information actually recorded is the time of arrival of each electron signal, conversion to electron energy is necessary, which involves calibration parameters. This is generally done using samples with known ionisation energies, as explained in Appendix 2.

7 Order and format of the spectra

Before starting on molecules we first include an atom, Xe, to illustrate the process of gathering and interpreting raw electron pair coincidence data. After that the molecules are presented in order of the number of atoms, then the number of valence electrons, much as in Herzberg's great compilation of electronic spectra [11]. For the larger polyatomic molecules, this scheme is sometimes departed from in order to group related molecules together.

The spectra have been taken by the TOF-PEPECO technique, using either a pulsed laboratory light source or synchrotron radiation. The resolution, understood as the width of a line for a single final state, depends on the electron energy, because for single electrons the measured linewidth is proportional to the energy itself over most of the useful energy range. The linewidth there is given roughly by $\Delta E = E/100$ or $\Delta E = E/50$, depending on whether the 5.6 m or 2 m apparatus was used. Although two electrons are involved in double ionisation, this form of relationship holds approximately here too, and the reduction in resolution at the higher photon and electron energies is very apparent in the spectra. Where possible, spectra covering the whole double ionisation range up to and a little beyond the triple ionisation energy are included, but these are not available for all molecules.

Tables of ionisation energies are included for most molecules, with uncertainties (error limits) implied by the quoted precision, but not given explicitly. Where energies are given to 0.01 eV precision, the implied uncertainty is ±0.02 eV. If several energies are given with the last figure 0 or 5 in the hundredths of an eV, the implied uncertainty is ±0.05 eV. Where energies are given with only one or no decimal places similar rules apply, but the likely accuracy can also be adjudged by examination of the spectra themselves.

The most common units, abbreviations and acronyms, which are used throughout the text are listed below, while a full list is included for easy reference on the inside end-covers of the book.

eV	electronvolt (1 eV = 8,066 cm^{-1}, 96.485 kJ mol^{-1}, 23.06 kcal mol^{-1})
meV	milli-electronvolt, 10^{-3} eV
Å	Ångström unit, 10^{-10} m, 0.1 nm or 100 pm
fs	femtosecond, 10^{-15} s
TOF-PEPECO	time-of-flight photoelectron–photoelectron coincidence
TPEsCO	threshold photoelectrons coincidence
DFKER	Doppler-free kinetic energy release

More acronyms identifying theoretical methods are given in Appendix 3.

References

The internet, with the help of a search engine, will provide more information on many of the topics covered here, and should probably be the reader's first port of call.

[1] P. Kruit and F.H. Read, *J. Phys. E: Scientific Instruments*, **16** (1983) 313.
[2] J.H.D. Eland et al., *Phys. Rev. Lett.*, **90** (2003) 053003.
[3] T.A. Carlson, *Photoelectron and Auger Spectroscopy*, Plenum Press, New York, 1975.
[4] J. Appell et al., *J. Phys. B: At. Mol. Phys.*, **6** (1973) 197.
[5] G. Dawber et al., *J. Phys. B: At. Mol. Opt. Phys.*, **27** (1994) 2191.
[6] M. Lundqvist et al., *Phys. Rev. Lett.*, **75** (1995) 1058.
[7] E.P. Månsson et al., *Nature Physics*, **10** (2016) 207.
[8] D.T. Attwood, *Soft X-rays and Extreme Ultraviolet Radiation*, Cambridge University Press, 1999.
[9] F.J. Duarte (editor), *Tunable Lasers Handbook*, Academic Press, New York, 1995.
[10] L.-S. Wang, H.-S. Cheng and J. Fan, *J. Chem. Phys.*, **102** (1995) 9480.
[11] G. Herzberg, *Electronic Spectra of Polyatomic Molecules*, Van Nostrand, New Jersey, 1966.

Chapter 1

Atomic xenon, Xe

1

Atomic xenon, Xe

To show how spectra are derived from the raw data, we first showcase an atom, in fact the very first atom to which the TOF-PEPECO technique was applied [1]. The lowest double ionisation energy of Xe is about 33 eV, so the He resonance line at 40.81 eV is easily energetic enough to cause double ionisation by absorption of a single photon. Since two electrons are emitted in the process, the raw data are inherently at least three-dimensional, as they must show intensity as a function of the energies of both electrons emitted in each double ionisation event. In principle the data could be highly multi-dimensional if we included the angles at which the electrons are emitted and the individual electron spins, but these are complications too far from the present purpose, besides being quite unknown at the present time. We represent the raw data as 'coincidence maps', with intensity indicated by colour and the electron characteristics on two perpendicular axes. The quantity actually measured is the flight time after a pulse of ionising light for each of the two electrons detected in coincidence, so the original map, Fig. 1.1, uses this time of flight (TOF) for the axes.

Figure 1.1 shows a family of nearly hyperbolic curves with definite ends at long flight times and gaps all along the leading diagonal. Each curve represents formation of a doubly ionised state at a definite double ionisation energy (DIE) above the atomic ground state, with emission of two electrons with energy sum equal to the photon energy minus the DIE:

$$E_1 + E_2 = h\nu - \text{DIE}.$$

The lines have definite ends because all electrons are accelerated slightly in the apparatus, so that even those that are created with zero initial energy arrive at a definite time. Using a simple formula with constants calibrated by measuring electrons of known energy, the times of flight can be converted into kinetic energies, thereby transforming the curves of Fig. 1.1 into the straight lines of Fig. 1.2. The formula relating times of flight TOF to energies E is

$$\text{TOF} = D/(E + E_0)^{1/2} + t_0$$

where D is the effective length of the flight path, E_0 is any extra energy given to or taken from the electrons by applied or stray electrical potentials and t_0 corrects for any time delays in the cables or detection circuits.

Double Photoionisation Spectra of Molecules. John H. D. Eland and Raimund Feifel.
© John H. D. Eland and Raimund Feifel, 2018. Published in 2018 by Oxford University Press.
DOI 10.1093/oso/9780198788980.001.0001

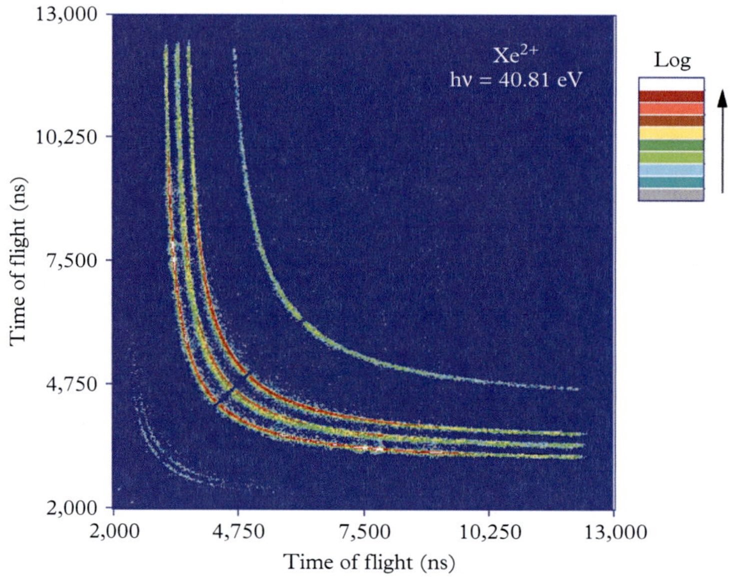

Figure 1.1 *Raw coincidence map for photoionisation of Xe by 40.81 eV photons, on a scale of electron flight times. Since one electron always arrives after the other, only half of the map shows actual data; it has been duplicated symmetrically about the main diagonal to make its form apparent. A gap appears on the diagonal because electrons are not recorded if they arrive exactly simultaneously.*

Instead of drawing the coincidence maps, we usually just add the electron energies together for each electron pair, and so draw the double ionisation spectrum, shown as Fig. 1.3. Each of the sharp peaks seen in the figure represents an electronic state of Xe^{2+}, as indicated. All five states in the lower spectrum arise from the same electron configuration, $4d^{10}5s^25p^4$, outside the deeper closed shells. The upper spectrum in the figure comes from double ionisation of Xe in Auger decay of the $^2D_{5/2}$ state from the configuration $4d^95s^25p^6$ which lies at 67.55 eV. In addition to the states seen with 40.81 eV photons, this Auger spectrum shows the $^1P_1$ state from the configuration $4d^{10}5s5p^5$ near 48 eV.

These spectra of Xe^{2+} introduce some features which are common to all the molecular spectra to be presented in the following pages. Firstly, each peak in the spectrum represents a state of the doubly charged ion. In Xe^{2+} these are all electronic states, but in molecules they can be vibrational levels within electronic states, or if vibrations are not resolved the peaks may turn into broad bands. Secondly, in ionisation from closed shells as in Xe, where all electrons are initially paired, the spectra contain both singlet and triplet spin states with comparable intensities, sometimes roughly in line with their statistical weights of 1:3. In Auger spectra, by contrast, triplet states are suppressed [2]. In Xe this effect is less marked than in molecules built from light atoms, because of

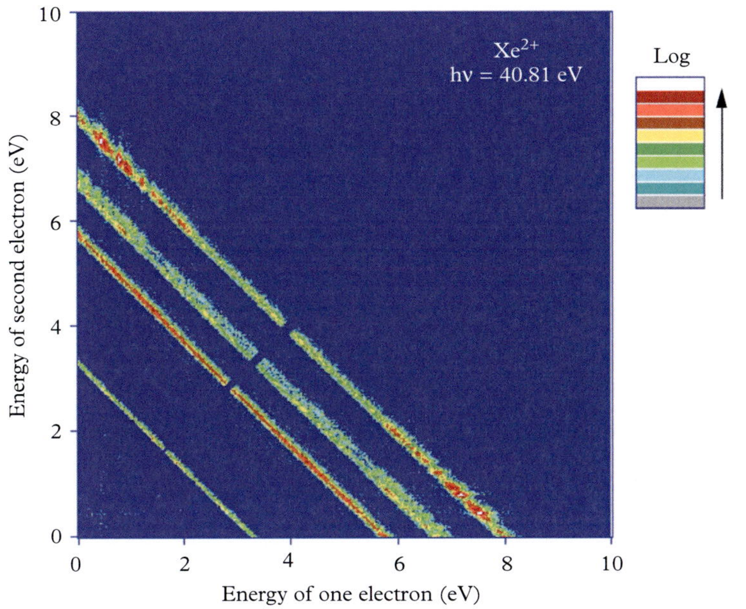

Figure 1.2 *The same data as in Fig. 1.1, drawn as a coincidence map in terms of electron energies for the main double ionisation processes.*

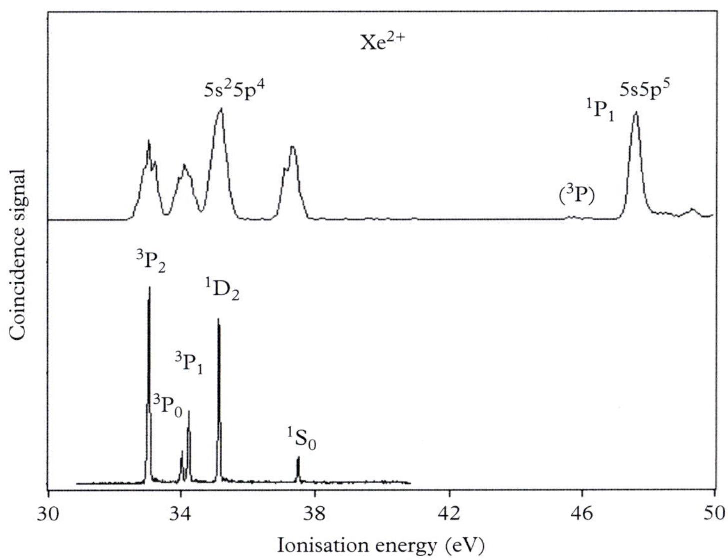

Figure 1.3 *Spectra of Xe^{2+} from photoionisation at 40.81 eV (lower trace) and from Auger decay of the $^2D_{5/2}$ state of Xe^+ (upper trace) formed by ionisation of a 4d electron at a photon energy of 90 eV.*

xenon's strong spin–orbit coupling. Thirdly, we see a big contrast in energy resolution between the photoionisation spectrum and the Auger spectrum, mainly due to instrumental effects which are common to all the spectra to be presented. The photoionisation spectrum was measured in the laboratory with a monochromatic source of 40.81 eV ionising light and a magnetic bottle apparatus 5.6 m long, the longest in the world at the time. The energy resolution is proportional to the length of the flight path in the bottle and also inversely proportional to the electron energy. For higher energy photons, we must use synchrotron radiation sources at large facilities, where space for apparatus is limited and the longest magnetic bottle we can use is 2 m long. Where the photon energy is high, photoelectron energies are high too, further increasing the energy bandwidth. In Auger spectra, another source of line broadening is the short lifetime τ of the initial state, according to the relationship

$$\tau \Delta E > h/2\pi.$$

If τ is expressed in femtoseconds and the energy in electronvolts, the constant term $h/2\pi$ in this expression is about 0.658 eV fs. The lifetimes of inner-shell hole states are usually in the low femtosecond range, producing broadening of the order of electronvolts. In principle, this lifetime broadening can be eliminated in the present technique if the photoelectron energy and the Auger electron energy are added together. The difference between this sum and the photon energy is then fixed by two quantum states of defined single energies (the neutral starting state and final dication state), and does not depend on the lifetime of the intermediate.

One last but important observation is to be made by examination of the data for Xe in the coincidence maps, shown in Figs. 1.1 and 1.2. The intensity along the distinct lines in these maps is not constant along the lines, but shows peaks on a relatively smooth or slowly varying background. This can be seen more clearly if we map the sum of the two electron energies against the lower energy of the two and then draw a three-dimensional projection on an expanded scale as in Fig. 1.4. The position of each horizontal line in this figure corresponds to the energy of a final state of Xe^{2+}, and the distance along the line shows the energy of the slower electron of the pair. The probability of double ionisation as a function of the slow electron's energy is indicated by colour coding or shown as a projection, and is clearly not uniform but has pronounced features, strongest in formation of the ground state of Xe^{2+}, the $^3P_2$ state.

Each peak in Fig. 1.4 shows that double photoionisation to the final state followed an *indirect* path rather than a *direct* one. Being indirect means that the process takes place in two stages, with intermediate formation of a singly charged intermediate:

$$m + h\nu \rightarrow m^{+*} + e_1^-; \quad m^{+*} \rightarrow m^{++} + e_2^-.$$

Because in Fig. 1.4 the maps are plotted against the energy of the slower electron, only one peak appears in the figure for each indirect pathway. In reality there are two peaks; one at the energy E_1 as shown, the other at a complementary energy $E_2 = h\nu - DIE - E_1$, where DIE is the ionisation energy to the final state identified by the energy sum $E_1 + E_2$.

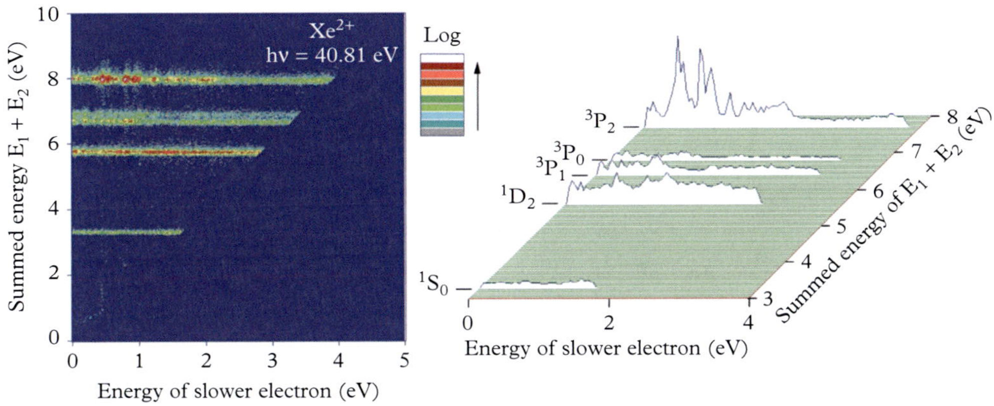

Figure 1.4 *The vertical axis in the coincidence map (left) is the sum of the two energies* $E_1 + E_2$ *and the horizontal axis is the lower energy of the two. In the isometric projection of the same data (right) the spacing between* $^3P_0$ *and* $^3P_1$ *has been exaggerated for clarity.*

The intermediate state m^{+*} may be called *superexcited*, because it lies in the continuum above the double ionisation limit, and the process in which it spontaneously ejects a second electron is called *autoionisation*. In a direct double ionisation process, by contrast, the two outgoing electrons share the available energy more or less randomly, forming a smooth and rather flat distribution underlying the peaks from any indirect processes. The direct contributions are seen in Fig. 1.4 to provide most of the intensity in formation of the higher energy final states, and contribute to the lowest state ($^3P_2$) when the electron energy E_1 is relatively high.

Indirect double photoionisation processes are very common in all atoms and molecules. In both atoms and molecules, the intermediate states can also give up their excess energy by photon emission or, in molecules only, by dissociation. The majority of autoionising intermediate states tend to be found only a little way above the lowest double ionisation energy (DIE), as here in Xe. More detailed characteristics of the direct and indirect photoionisation processes are discussed in Appendix 1.

The attentive reader may have noticed that there are three weak 'hyperbolae' in Fig. 1.1 which have been ignored until now. These lines arise from a process in which a single pulse of light ionises two separate Xe atoms producing two Xe^+ ions. Each atomic ion can be formed in one of the two spin–orbit split levels, $^2P_{3/2}$ and $^2P_{1/2}$, with statistical weights of 2:1. The three possibilities $^2P_{3/2}+^2P_{3/2}$, $^2P_{3/2}+^2P_{1/2}$ and $^2P_{1/2}+^2P_{1/2}$ account for the three hyperbolae, whose relative intensities follow the combined statistical weights. From this analysis, we know what states have been formed, but not how they were formed. From the apparent DIE of 24.2 eV for the lowest hyperbola, we know that two Xe^+ ions have been formed in the ground state by a single photon, and from the continuous form of the hyperbolae it is clear that excess energy has been shared more or less randomly between the two ejected electrons. The mechanism must be primary ionisation of one Xe atom by the photon followed by ionisation of a second atom by the outgoing electron in an inelastic collision. As the primary electron is of a known energy, controllable by selection of the photon energy, this phenomenon could be used to study electron impact ionisation in a new way. Because two atoms are required, the intensity of the process depends on the square of the target gas pressure, and can be reduced to negligible importance at low pressure. The process is also recognisable by the impossibly low apparent double ionisation energies it produces.

To conclude this introductory chapter, we summarise some basic features of double photoionisation and of the spectra presented in this book, exemplified by Xe.

- The raw data are the times of flight of two electrons in coincidence from single photoionisation events.
- The energy resolution is good for low electron energies; less good, the higher the energy.
- Ionisation may be *direct*, producing a continuous spread of electron energies or *indirect*, producing peaks in the energy distributions.
- An extreme indirect process is Auger decay from an inner-shell hole state. Here singlet states of the final ion are strongly favoured if the starting atom or molecule had all electrons paired. In photoionisation at energies below the binding energies of inner shells, both singlet and triplet states are populated efficiently.

References

[1] J.H.D. Eland et al., *Phys. Rev. Lett.*, **90** (2003) 053003.
[2] H. Ågren, *J. Chem. Phys.*, 75 (1981) 1267.

Chapter 2

Diatomic molecules

2

Diatomic molecules

The first diatomic molecule in a sequence based on the number of electrons should be H_2, but because there are no bound states of the dication, only a close combination of two bare protons, we do not include its spectrum here. Double photoionisation of hydrogen to the repulsive state of $H^+ + H^+$ has been studied both experimentally and theoretically [1–4]. Because of energy conservation and the small number of particles, the spectrum can be seen in the kinetic energies of the protons as well as in the electron pairs, and is a smooth bump. To a first approximation it is a reflection of the ground state wavefunction of H_2 in the potential curve for Coulomb repulsion.

After H_2 should come the diatomic hydrides, starting with HF. Neither we nor any other group has dared to put this extremely corrosive and chemically aggressive gas into a spectrometer to measure its double photoionisation spectrum. From the electron binding energies revealed in its photoelectron spectrum [5] we can estimate that its lowest double ionisation energy will be of the order of 45 eV and so beyond the range of HeIIα photoionisation.

For the next molecule, HCl, the double ionisation energies (DIEs) are known experimentally from its TPEsCO spectrum [6] but not yet from TOF-PEPECO. In general form the TPEsCO spectrum closely resembles the TOF-PEPECO spectra of HBr and HI, presented in this chapter. Of the heavier diatomic molecules we are also short of F_2, Cl_2 and various diatomic interhalogen compounds, for the self-same reason, that they have a very deleterious effect on sensitive apparatus. The TPEsCO spectrum of Cl_2 is also available [6].

For all molecules, including diatomics, there are more possible routes to double ionisation than the single-step removal of two electrons. For a molecule AB the lowest possible energy of the bound molecular dication AB^{2+} is almost invariably higher, often considerably higher, than the minimum energy of the separated charged products $A^+ + B^+$. The only exceptions to this are highly exotic species involving rare gas or very heavy atoms. So double ionisation can take place by dissociation first and ionisation afterwards, at energies either above or below the molecular DIE, by at least two mechanisms:

$$AB + h\nu \rightarrow A^+ + e_1^- + B^*$$

$$B^* \rightarrow B^+ + e_2^-$$

Double Photoionisation Spectra of Molecules. John H. D. Eland and Raimund Feifel.
© John H. D. Eland and Raimund Feifel, 2018. Published in 2018 by Oxford University Press.
DOI 10.1093/oso/9780198788980.001.0001

or alternatively

$$AB + h\nu \rightarrow A^* + B^*$$

$$A^* \rightarrow A^+ + e_1^-$$

$$B^* \rightarrow B^+ + e_2^-$$

The step in which a superexcited atom A^* spontaneously emits an electron is called *autoionisation* and may take place some time after the formation of A^*, allowing the species to separate. The separation distances cover a range governed by the lifetime of A^*, so the final energy is also smeared out by the range of Coulomb repulsion energies in the separated ions. The result is that these processes are not easily seen in double ionisation spectra, but are highly visible in the coincidence maps. They are found to be most prevalent in small molecules containing terminal oxygen atoms, but have also been found in compounds of nitrogen, the halogens, sulphur and other atoms. The autoionising species are usually atoms, but can also be small molecules.

Double ionisation of most of the easily experimentally accessible diatomic molecules has been studied by several techniques, including Auger spectroscopy, double electron transfer, kinetic energy release and high-level theory. In what follows we give references to selected and mainly more recent publications which provide a key to earlier work.

References

[1] W. Vanroose et al., *Phys. Rev.* A, **74** (2006) 052702.
[2] R. Dörner et al., *Phys. Rev. Lett.*, **81** (1998) 5776.
[3] H. Kossman et al., *Phys. Rev. Lett.*, **63** (1989) 2040.
[4] H. Le Rouzo, *Phys. Rev.* A, **37** (1988) 1512.
[5] J. Berkowitz, *Chem. Phys. Lett.*, **11** (1971) 21.
[6] A. McConkey et al., *J. Phys. B: At. Mol. Opt. Phys.*, **27** (1994) 271.

2.1 Hydrogen bromide, HBr

The electron configuration of HBr in its outermost shells can be written as

$$3d^{10}4s\sigma^{2}4p\sigma^{2}4p\pi^{4}\ldots\ldots\ldots\ldots{}^{1}\Sigma_{0}^{+}.$$

The 3d shell is at a binding energy of about 70 eV, so it cannot be broached using HeII light, while any contribution of the 4sσ electrons at around 26 eV to double ionisation has not yet been observed. The photoelectron spectrum [1, 2] and the threshold photoelectron spectrum [3] show bands for ionisation from the 4pπ orbital and 4pσ orbital at about 11.8 and 15.5 eV respectively. Interpreting these ionisation energies as orbital energies in the spirit of Koopmans' theorem we can estimate the double ionisation energies (DIEs) in an approximate way as sums of two orbital energies plus a Coulomb repulsion energy, as follows:

$$\mathrm{IE}\left(4p\pi^{-2}\right) = 2 \times 11.8 + 14.4/1.4 = 33.9\,\mathrm{eV};$$

$$\mathrm{IE}\left(4p\pi^{-1}, 4p\sigma^{-1}\right) = 11.8 + 15.5 + 14.4/1.4 = 37.5\,\mathrm{eV};$$

$$\mathrm{IE}\left(4p\sigma^{-2}\right) = 2 \times 15.5 + 14.4/1.4 = 41.3\,\mathrm{eV}.$$

In these calculations, 14.4 is the Coulomb energy in eV for unit charges 1 Å apart and the charge separation is crudely estimated as equal to the normal H—Br bond distance. Such very simple calculations help to interpret many double photoionisation spectra including that of HBr shown here.

In the spectrum of Fig. 2.1.1 there are three bands with vibrational structure, centred overall at about 34 eV. These must represent the three states which arise from the $4p\pi^2$ configuration, $X^3\Sigma^-$, $a^1\Delta$ and $b^1\Sigma^+$, as marked on the spectrum. It is reasonable to expect that these states should be bound and show vibrational structure since the

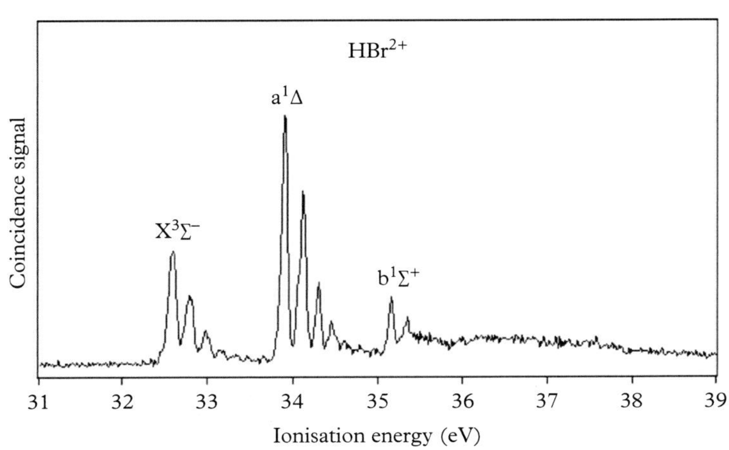

Figure 2.1.1 *TOF-PEPECO spectrum of HBr taken using HeIIα 40.81 eV light and, like all the HeII spectra presented in this chapter and later, with the 5.6 m magnetic bottle mentioned in the Introduction chapter.*

$4p\pi$ electrons are non-bonding 'lone pairs' in chemical parlance, and their loss will not cause immediate dissociation. The $4p\sigma$ electrons, on the other hand, are strongly bonding; loss of even one of them causes dissociation, as seen by break-off of vibrational structure in the photoelectron spectrum and in fluorescence emission [1–3]. Thus we can interpret the very broad unstructured band in Fig. 2.1.1 centred around 37 eV as an unresolvable blend of the two electronic states $^3\Pi$ and $^1\Pi$ which arise from the configuration $4p\sigma^1 4p\pi^3$. The $^1\Sigma^+$ state expected to arise from removal of two electrons from $4p\sigma$ will undoubtedly give rise to an equally broad and structureless band, probably at an energy inaccessible with 40.81 eV light.

A TOF-PEPECO spectrum of HBr^{2+}, giving a little more detail on the vibrational structure but with overoptimistic absolute energy precision has been published [4] and the DIEs have also been determined by the TPEsCO technique [5, 6] and by Auger spectroscopy [7, 8], all in good agreement with the energies given in the list that follows below. The high-resolution Auger spectrum [7, 8] also contains broad bands corresponding to ionisation energies of 37 and 43 eV, which can be attributed to the $4p\sigma^1 4p\pi^{-1}$ and $4p\sigma^{-2}$ transitions respectively. The Auger decay in HBr is particularly interesting because it is in competition with rapid dissociation to excited bromine atoms, which later autoionise [9]. Theoretical calculations of the DIEs to the low-lying states have also been carried out by several techniques; the most recent work [10] gives references to the earlier calculations.

States of HBr^{2+}

State	DIE (eV)*	Ionisation	Vibration (meV)
$X^3\Sigma^-$	**32.60**	$4p\pi^{-2}$	206 ± 5
$a^1\Delta$	**33.90**	$4p\pi^{-2}$	209 ± 5
$b^1\Sigma^+$	**35.17**	$4p\pi^{-2}$	200 ± 8
$^1\Pi, {}^3\Pi$	37 ± 1	$4p\sigma^{-1}4p\pi^{-1}$	

* Adiabatic DIEs are printed in **bold type** here and elsewhere.

References

[1] B.S. Schneider and A.L. Smith in D.A. Shirley (Ed.), *Electron Spectroscopy*, North Holland, Amsterdam, 1972.

[2] C.E. Brion and P. Crowley, *J. Electron Spectrosc. Related Phenom.*, **11** (1977) 399.

[3] A.J. Yencha et al., *Chem. Phys.*, **238** (1998) 133.

[4] J.H.D. Eland, *Chem. Phys.*, **294** (2003) 171.

[5] A.J. Yencha et al., *Chem. Phys. Lett.*, **381** (2003) 609.

[6] M. Alagia et al., *J. Chem. Phys.*, **120** (2004) 6980.

[7] B. Wannberg et al., *Chem. Phys.*, **133** (1989) 281.

[8] R. Püttner et al., *Phys. Rev. A*, **59** (1999) 4438.

[9] P. Morin and I. Nenner, *Phys. Rev. Lett.*, **56** (1986) 1913.

[10] L. Streit, F.B.C. Machado and R. Custodio, *Chem. Phys. Lett.*, **506** (2011) 22.

2.2 Hydrogen iodide, HI

The double ionisation spectrum of HI was first measured and interpreted by the TPEsCO method [1] and proved to be very similar to the spectra of HCl^{2+} and HBr^{2+}. The TOF-PEPECO spectrum shown here is in excellent agreement with the TPEsCO results on the vibrationally resolved part of the spectrum and it also extends to higher energy, showing further bands. The electron configuration of the outermost shells is

$$4d^{10}5s\sigma^2 5p\sigma^2 5p\pi^4 \ldots\ldots\ldots {}^1\Sigma^+.$$

The same simple calculations as used for HBr based on the HI photoelectron spectrum [2] and an approximate bond length of 1.6 Å predict the bands from $5p\pi^{-2}$ ionisation at around 30 eV, those from $5p\sigma^{-1}5p\pi^{-1}$ near 34 eV, one band from $5p\sigma^{-2}$ ionisation at 37 eV and bands from $5s\sigma^{-1}5p\pi^{-1}$ at 42 eV. Other states are expected to lie at even higher energies, where single ionisation from the 4d shell of iodine at around 51 eV will also enable Auger decay.

The TOF-PEPECO spectrum shown as Fig. 2.2.1 contains the expected three bands from $5p\pi^{-2}$ ionisation, all with resolved vibrational structure, and a broad unstructured band with its peak near 35 eV which we attribute to the $^3\Pi$ and $^1\Pi$ states from $5p\sigma^{-1}5p\pi^{-1}$ ionisation. In another TOF-PEPECO spectrum [3] taken with 51 eV (HeIIγ) light, a further very broad and apparently structureless band appears at around 38 eV; this is presumably the singlet state from $5p\sigma^{-2}$ ionisation.

A notable feature of the spectrum in Fig. 2.2.1 is that in the $^3\Sigma^-$ band the second resolved peak is more intense than the first, which is the opposite of the situation in HBr^{2+} (and HCl^{2+}). Yencha et al. [1] pointed out that this difference arises from the very strong spin-orbit coupling in iodine, which splits the $^3\Sigma^-$ state into two components $^3\Sigma^-{}_0$ and $^3\Sigma^-{}_{\pm 1}$, with a spacing close to the vibrational interval. The relative intensities

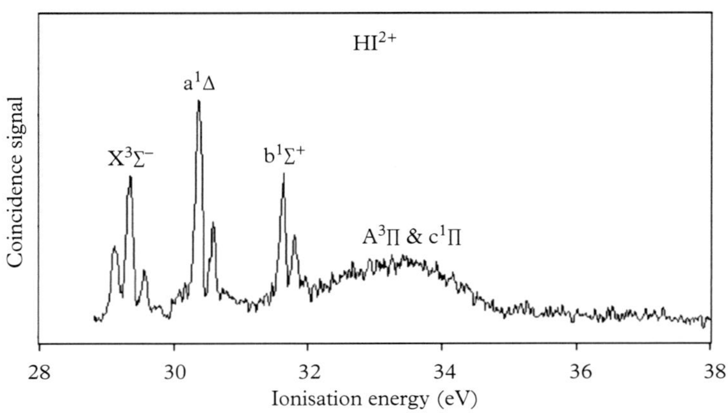

Figure 2.2.1 *TOF-PEPECO spectrum of HI taken with HeIIα 40.81 eV light.*

of the two components are evidently in line with their statistical weights of 1:3, as can be deduced directly from their intensities in this spectrum. The TPEsCO spectrum shows somewhat different though similar relative intensities, probably because of unquantified effects of autoionisation. The vibrational constants and potential energy curves for HI^{2+} in its resolved states were calculated by Yencha et al. [1], and also using a different theoretical method by Ghosh et al. [4].

States of HI^{2+}

State	DIE (eV)*	Ionisation	Vibration (meV)
$X^3\Sigma_0^-$	**29.14**	$5p\pi^{-2}$	216 ± 10
$X^3\Sigma_1^-$	**29.34**	$5p\pi^{-2}$	216 ± 10
$a^1\Delta$	**30.37**	$5p\pi^{-2}$	223 ± 10
$b^1\Sigma^+$	**31.62**	$5p\pi^{-2}$	183 ± 10
$A^3\Pi$, $c^1\Pi$	33.6 ± 1	$5p\sigma^{-1}5p\pi^{-1}$	
$^1\Sigma^+$	ca 38	$5p\sigma^{-2}$	

* Adiabatic DIEs are printed in **bold type** here and elsewhere.

References

[1] A.J. Yencha et al., *Chem. Phys.*, **303** (2004) 348.

[2] M.Y. Adam et al., *Chem. Phys.*, **164** (1992) 123.

[3] A. Pilcher-Clayton and J.H.D. Eland, *J. Electron Spectrosc. Related Phenom.*, **142** (2005) 313.

[4] D. Ghosh, N.C. Bera and A.K. Das, *Chem. Phys. Lett.*, **461** (2008) 348.

2.3 Nitrogen, N_2

Nitrogen, as our first diatomic molecule containing two heavy atoms, has a richer electron configuration and a much richer set of states in its doubly charged form than the diatomic hydrides. Its configuration outside the inner 1s shells can be written as

$$2s\sigma_g{}^2 2s\sigma_u{}^2 2p\pi_u{}^4 2p\sigma_g{}^2 \ldots\ldots\ldots\ {}^1\Sigma_g{}^+$$

where the atomic orbital basis of the molecular orbitals is explicit. Alternatively, in the conventional form where the orbitals of each symmetry are counted and numbered from the deepest, the orbitals including the 1s shell are

$$1\sigma_g{}^2 1\sigma_u{}^2 2\sigma_g{}^2 2\sigma_u{}^2 1\pi_u{}^4 3\sigma_g{}^2 \ldots\ldots\ldots\ {}^1\Sigma_g{}^+.$$

By removing two electrons from the valence and inner valence orbitals in all possible combinations and allowing for the spin (electrostatic) splittings, we can generate 18 distinct electronic states which should lie in the energy range below 90 eV. Simple Koopmans-inspired calculations of the form used for HBr and HI are of little use here, partly because the orbital energies revealed in the photoelectron spectrum are rather close together, and also because the very concept of one-electron wavefunctions (orbitals) begins to lose its validity. In other words, electron correlation becomes important for the character and energy ordering of the electronic states. We can only give the dominant electron configurations of the observed states, aided by high-level theory. Most of the states seen in Fig. 2.3.1 and Figs 2.3.2 also have extended vibrational structure or broad bands, and many overlap one another.

The spectra in Fig. 2.3.1 illustrate a common feature of these and many spectra; high resolution favours sharp structures, while broad features are brought out more strongly by low resolution. The lower spectrum was taken under regular conditions of modest (2% or 1 in 50) resolution, whereas for the upper spectrum every effort was made to achieve the best possible resolution, about 0.8% or 1 in 125. The sharp vibrational levels are best seen at high resolution while the mostly unbound $d^3\Pi_g$ state is seen far better at lower resolution. In Fig. 2.3.2, measured with 70 eV photons, the energy bandwidth is wide because of the higher electron energies and the shorter (2 m) flight path, and this effect is even more marked.

The assignment of the states in these spectra is based on the analysis of Ahmad et al. [1] which also included a TPEsCO spectrum extending up to 54 eV. A well-resolved spectrum of the dissociative states of N_2^{2+} was deduced by Lundqvist et al. [2] from fragment ion kinetic energy releases. The spectrum from double charge transfer (DCT) to H^+ by Furuhashi et al. [3] shows the first singlet state, $X^1\Sigma_g{}^+$, with clear vibrational structure, and the singlet A state. The Auger spectrum [4, 5] also emphasises singlet states, at low resolution. More complete references to earlier experimental work by several techniques are given by Lundqvist et al. [2]. Rather surprisingly the $D^1\Sigma_u{}^+$ state near 51 eV emits fluorescence from its v=0 and v=1 levels in falling down to the ground state, and the emission spectrum has been analysed in detail [6, 7]. There have been theoretical calculations of the potential energy curves for the dication relating to experimental

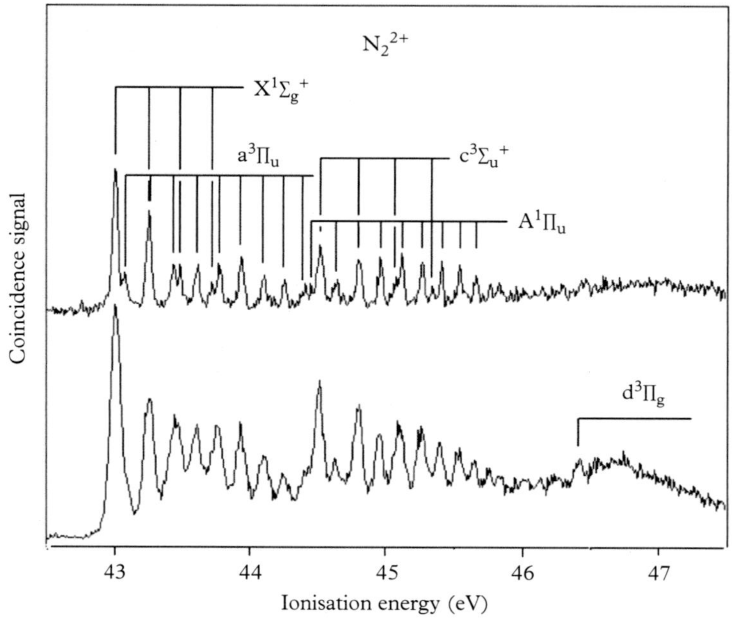

Figure 2.3.1 *TOF-PEPECO spectra of N_2 taken with HeIIβ light at 48.37 eV photon energy using two different instrumental resolution conditions (see text).*

works [3, 4, 8], and those of Senekowitsch et al. [8], which are particularly helpful for interpretation of the spectra. The triple ionisation energy of N_2 is estimated as near 90 eV, but as double ionisation spectra taken at 100 eV [9] show only weak features beyond those seen in Fig. 2.3.2, we believe that the figure is an essentially complete double photoionisation spectrum. Most of the states that can be formed by removing two electrons from the valence orbitals of N_2 have been identified in the TOF-PEPECO spectra, with the notable exception of $^3\Sigma_g^-$, $^1\Delta_g$ and $^1\Sigma_g^+$, nominally from removal of two electrons from the $2p\pi_u$ orbital. Because of the strongly bonding character of this orbital, these states will have repulsive potential energy surfaces or at most shallow potential wells and an extended band structure which may be concealed in the background. Peaks attributed to the first two of these missing states are seen, however, in the TPEsCO spectrum; the reason for their absence from the TOF-PEPECO spectra is not known. Another important source of information on the states of N_2^{2+} is the Auger spectrum, remeasured with vibrational resolution by Püttner et al. [10], who confirm the identifications given by Ahmad et al. [1]. The state tentatively identified as $^3\Delta_u$ near 53 eV and marked as such in Fig. 2.3.2 may in fact be an expected $^1\Sigma_u^+$ state from $2s\sigma_u^2$ ionisation as proposed by Ågren [11] in theoretical interpretation of the Auger spectrum.

The high energy bands mentioned in the caption to Fig. 2.3.2 arise from double ionisations involving the inner valence orbitals, whose ionisation energies span a range from 25 to 40 eV in the photoelectron spectrum. The dication states are presumably based on configurations with two vacancies in molecular orbitals arising from atomic N2s orbitals, probably with admixture of other configurations.

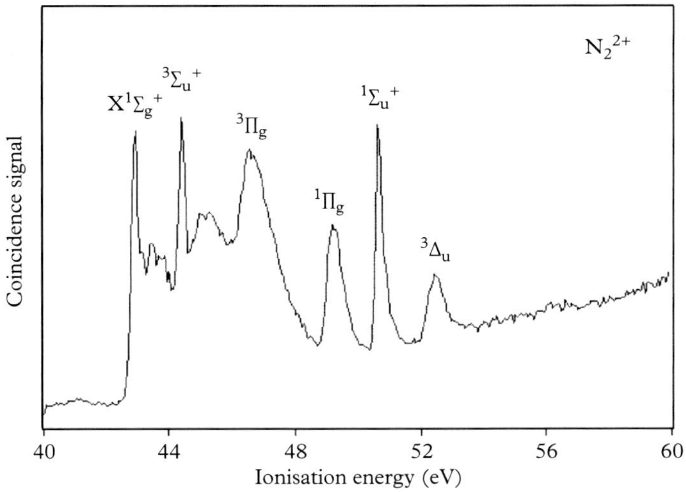

Figure 2.3.2 *Wide-range TOF-PEPECO spectrum of nitrogen taken at 70.0 eV using synchrotron radiation. A spectrum taken at 100 eV photon energy [9] shows a weak feature at 56 eV (faintly visible above) and two further very broad bands at 68 and 72 eV.*

States of N_2^{2+} (some not shown in figures)

State	DIE (eV, bold if adiabatic)	Major orbitals
$X^1\Sigma_g^+$	**42.88**	$2p\sigma_g^{-2}$
$a^3\Pi_u$	**42.98**	$2p\pi_u^{-1}2p\sigma_g^{-1}$
$b^3\Sigma_g^-$	43.7	$2p\pi_u^{-2}$
$A^1\Pi_u$	**44.4**	$2p\pi_u^{-1}2p\sigma_g^{-1}$
$^1\Delta_g$(?)	44.4	$2p\pi_u^{-2}$
$c^3\Sigma_u^+$	**44.5**	$2p\sigma_g^{-1}2s\sigma_u^{-1}$
$d^3\Pi_g$	46.3	$2p\pi_u^{-1}2p\sigma_u^{-1}$
$B^1\Pi_g$	49.5	$2p\pi_u^{-1}2p\sigma_u^{-1}$
$D^1\Sigma_u^+$	**50.8**	$2p\sigma_g^{-1}2s\sigma_u^{-1}$
‘$^3\Delta_u$’	52.7	
$^{1,3}\Sigma_u^+$	68	$2s\sigma_g^{-1}2s\sigma_u^{-1}$
$^1\Sigma_g^+$	72	$2s\sigma_g^{-2}$

References

[1] M. Ahmad et al., *J. Phys. B: At. Mol. Opt. Phys.*, **39** (2006) 3599.
[2] M. Lundqvist et al., *J. Phys. B: At. Mol. Opt. Phys.*, **29** (1996) 1489.
[3] O. Furuhashi et al., *Chem. Phys. Lett.*, **337** (2001) 97.
[4] W.E. Moddeman et al., *J. Chem. Phys.*, **55** (1971) 2317.
[5] S. Svensson et al., *J. Phys. B: At. Mol. Opt. Phys.*, **25** (1992) 135.
[6] D. Cossart, F. Launay, J.M. Robbe and G.J. Gandara, *Mol. Spectrosc.*, **113** (1985) 142.
[7] A. Ehresmann et al., *J. Phys. B: At. Mol. Opt. Phys.*, **33** (2000) 473.
[8] J. Senekowitsch, S. O'Neil, P. Knowles and H.-J. Werner, *J. Phys. Chem.*, **95** (1991) 2125.
[9] J.H.D. Eland et al., *Chem. Phys. Lett.*, **646** (2016) 31.
[10] R. Püttner et al., *J. Phys. B: At. Mol. Opt. Phys.*, **41** (2008) 141001.
[11] H. Ågren, *J. Chem. Phys.*, 75 (1981) 1267.

2.4 Carbon monoxide, CO

Carbon monoxide is isoelectronic with nitrogen in the valence shell, with the electron configuration

$$\mathrm{O1s^2C1s^23\sigma^24\sigma^21\pi^45\sigma^2}\,\ldots\ldots\ldots\ldots\ldots\ldots\,{}^1\Sigma^+.$$

The 3σ and 4σ orbitals are built mainly from C2s and O2s atomic orbitals and the corresponding photoelectron bands span a wide range of energies. There are many states involving ionisation from these orbitals because the orbital model, which neglects electron correlation and configuration mixing, is inadequate in the inner valence region and 3h-1p configurations play a major part. The part of the TOF-PEPECO double photoionisation spectrum which exhibits vibrational structure is shown in Fig. 2.4.1.

In the figure, five resolved states are identified, all of which are also seen in the related TPEsCO spectrum [1]. The singlet states are also seen in the vibrationally resolved DCT spectrum [2] and the dissociative levels in the Doppler-free kinetic energy release spectrum [3]. All the 'pseudo-stable' states which show vibrational structure have been characterised in some detail by Püttner et al. [4] on the basis of well-resolved

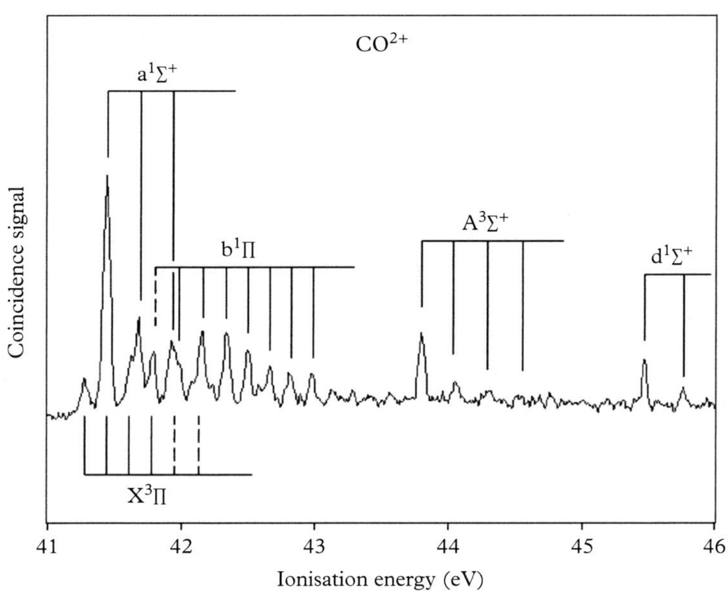

Figure 2.4.1 *TOF-PEPECO spectrum of CO measured using HeIIβ 48.37 eV light. A weak unstructured band is also seen at 47.5 eV, particularly under low resolution conditions. Some of the background signal in this spectrum comes from dissociative double ionisation.*

Auger spectra. The unstructured band at 47.5 eV, mentioned in the caption, is interpreted as the inner repulsive wall of the $^1\Delta$ state from $1\pi^{-2}$ ionisation. The only information on higher energy states available at present comes from Auger spectra taken above the carbon and oxygen 1s edges, measured by Moddeman et al. [5] and analysed by Ågren and Siegbahn [6]. A more remarkable feature, not apparent from the well-resolved spectrum in Fig. 2.4.1 where its contribution has been suppressed, is that double ionisation also occurs at photon energies below the molecular double ionisation energy of 41.3 eV. The first evidence for this process in CO came from electron spectra [7], and it was later characterised more fully using the TOF-PEPECO and related techniques [8].

Figure 2.4.2 shows the dissociative double ionisation of CO at a photon energy of 40.81 eV, which is below the molecular double ionisation threshold. The process involved is initial single ionisation to a highly excited state of CO^+; this then dissociates to C^+ and a superexcited state of neutral atomic oxygen, O^*, which later autoionises. The products are C^+ and O^+, formed mainly in their ground states and with only a little

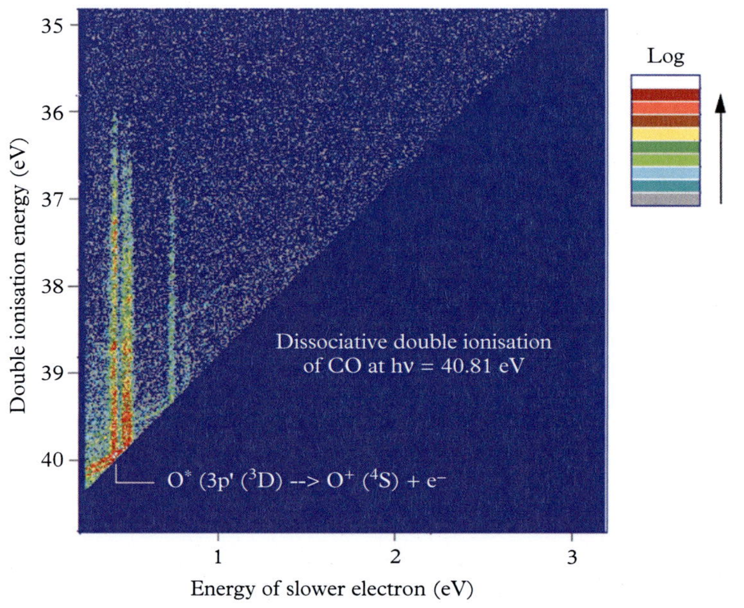

Figure 2.4.2 *Double photoionisation of CO at a photon energy below the minimum where bound CO^{2+} can be created. The distinct vertical lines show processes where an intermediate state of CO^+ is created first; this then dissociates to $C^+ + O^*$, where the O^* atom is in a metastable 'superexcited' state above its lowest ionisation threshold. The O^* atom autoionises in a final step, forming one of the accessible states of O^+ and releasing an electron of fixed energy.*

kinetic energy, showing that the final autoionisation takes place after the atoms have separated to a significant distance. The calculation of energies and distances in this instance is illustrated in Appendix 3, under the heading 'Thermochemical calculations'. Most of the intermediate CO^+ states involved can be identified in the high-resolution inner-valence photoelectron spectrum taken by Baltzer et al. [9]. The autoionising states of O^* which appear here also appear in coincidence maps for several other small molecules containing O atoms, and many have been identified [10]; one particular autoionisation pathway is identified in Fig. 2.4.2. Carbon monoxide is by no means the only molecule for which this sort of dissociative double photoionisation is important, as we shall see, and oxygen is not the only atom to autoionise in the same way.

When the spectrum of CO^{2+} is extracted without suppression of atomic autoionisation, dissociative double ionisation below the molecular threshold is very apparent, as seen in Fig. 2.4.3 where a photon energy of 80 eV has been used. The states already identified in Fig. 2.4.1 all reappear at the higher photon energy, and two broad bands are visible in Fig. 2.4.3 at around 48 and 51 eV. These bands probably comprise more than one state in each case, but from the Auger spectra [6] listed by Ågren [11] we tentatively identify significant components as the $^1\Delta$ and $^1\Sigma^+$ states from $1\pi^{-2}$ ionisation. The onset of $C^+ + O^+$ formation is seen to occur at or just above 36 eV, which is the thermodynamic threshold for formation of these atomic ions in their ground states, as discussed in Appendix 3. This means that final ionisation to create the ion pair must happen when the atomic species are far apart, as otherwise the Coulomb repulsion between them would increase their joint kinetic energies and raise the onset energy.

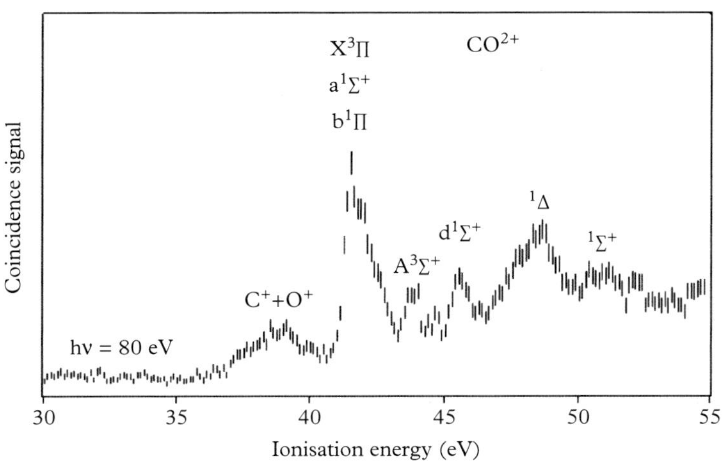

Figure 2.4.3 *TOF-PEPECO spectrum of CO at 80 eV photon energy measured using the 2 m magnetic bottle with monochromatic synchrotron light. The resolution is much poorer than in Fig. 2.4.1 because of the higher electron energies and the shorter flight path. The data points are shown as error bars, each with a length corresponding to two standard deviations.*

States of CO^{2+}

State	DIE (eV, bold if adiabatic)	Orbitals
$X^3\Pi$	**41.28**	$5\sigma^{-1}1\pi^{-1}$
$a^1\Sigma^+$	**41.44**	$5\sigma^{-2}$
$b^1\Pi$	**41.8**	$5\sigma^{-1}1\pi^{-1}$
$A^3\Sigma^+$	**43.80**	$4\sigma^{-1}5\sigma^{-1}$
$d^1\Sigma^+$	**45.46**	$4\sigma^{-1}5\sigma^{-1}$
$^1\Delta$	48	$1\pi^{-2}$
$^1\Sigma^+$	51	$1\pi^{-2}$

References

[1] J.H.D. Eland et al., *J. Phys. B: At. Mol. Opt. Phys.*, **37** (2004) 3197.
[2] O. Furuhashi et al., *Chem. Phys. Lett.*, **337** (2001) 97.
[3] M. Lundqvist et al., *Phys. Rev. Lett.*, **75** (1995) 1058.
[4] R. Püttner et al., *Chem. Phys. Lett.*, **445** (2007) 6.
[5] W.E. Moddeman et al., *J. Chem. Phys.*, **55** (1971) 2317.
[6] H. Ågren and H. Siegbahn, *Chem. Phys. Lett.*, **72** (1980) 498.
[7] U. Becker et al., *Phys. Rev.* A, **45** (1992) R1295.
[8] Y. Hikosaka and J.H.D. Eland, *Chem. Phys.*, **299** (2004) 147.
[9] P. Baltzer et al., *J. Phys. B: At. Mol. Opt. Phys.*, **27** (1994) 4915.
[10] R. Feifel, J.H.D. Eland and D. Edvardsson, *J. Chem. Phys.*, **122** (2005) 144308.
[11] H. Ågren, *J. Chem. Phys.*, **75** (1981) 1267.

2.5 Nitric oxide, NO

Nitric oxide has one more electron than N_2 or CO which goes into the antibonding π-orbital, so its electron configuration is

$$O1s^2N1s^23\sigma^24\sigma^21\pi^45\sigma^22\pi^1 \ldots\ldots\ldots\ldots\ldots\ldots ^2\Pi_{1/2,3/2}.$$

The 3σ and 4σ orbitals are made mainly from O2s and N2s atomic orbitals with binding energies around 40 and 22 eV according to the photoelectron spectrum [1], while the outer orbitals 1π, 5σ and 2π have binding energies of around 17, 16 and 10 eV respectively. Although there are at least seven ways of removing two electrons from these orbitals which might be expected to give double ionisation below 50 eV, only three resolved bands are seen in the TOF-PEPECO double ionisation spectrum [2] within this range, as shown in Fig. 2.5.1. Spin-orbit splitting in the ground state of NO is only 15 meV and is not resolved in these spectra.

The same vibrationally resolved states are also seen clearly in the TPEsCO spectrum of Dawber et al. [3], who carried out a vibrational analysis. The dissociative B state of NO^{2+} was previously seen by the Doppler-free kinetic energy release (DFKER) technique [4] and both that state and the A state are seen in optical emission [5, 6], the B state at high resolution [7]. The DFKER spectrum also revealed two more quasi-bound

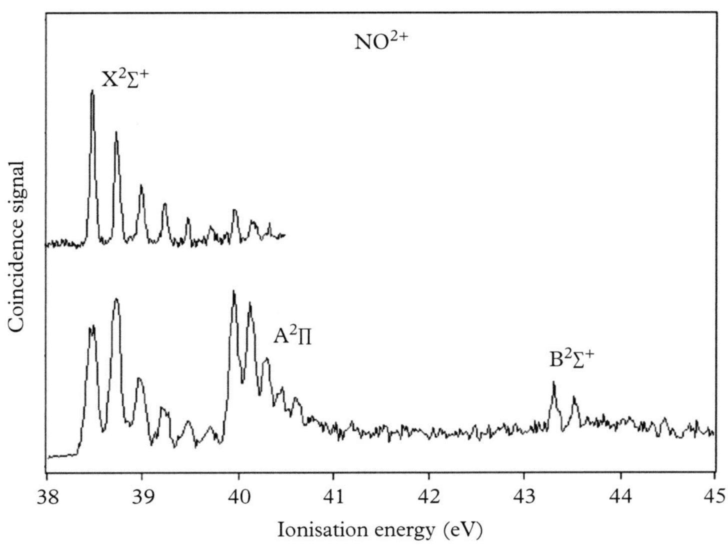

Figure 2.5.1 *TOF-PEPECO spectra of NO measured using HeIIα 40.81 eV and HeIIβ 48.37 eV light. A spectrum taken with HeIIγ 51.02 eV light [2] also showed a weak and apparently unstructured band at about 47 eV. Effects of dissociative double ionisation are not shown in these spectra, which emphasise molecular double ionisation.*

states supporting short-lived vibrational levels at 47.37 and 48.93 eV, which were identified with the aid of extensive theoretical calculations [4]. Extensive calculations on the doublet states were also carried out in connection with the Auger spectra from $N1s^{-1}$ and $O1s^{-1}$ ionisation [8]. The Auger spectra themselves show the X, A and B states and also intense bands in the range 47–50 eV and at higher energies. Some interpretation of these bands was proposed [8] but remains speculative for the moment.

Dissociative double photoionisation proceeding by autoionisation of atomic O* is as extensive in NO as in CO [2] and again extends to energies below the molecular double ionisation threshold.

States of NO^{2+}

State	DIE (eV, bold if adiabatic)	Orbitals ionised
$X^2\Sigma^+$	**38.46**	$5\sigma^{-1}2\pi^{-1}$
$A^2\Pi$	**39.76**	$1\pi^{-1}2\pi^{-1}$
$B^2\Sigma^+$	**43.30**	$4\sigma^{-1}2\pi^{-1}$
$C^2\Sigma^+$	**47.37**	$5\sigma^{-1}1\pi^{-1}$
$2^4\Pi$	**48.93**	$1\pi^{-2}$

References

[1] H.-J. Freund, H. Kossman and V. Schmidt, *Chem. Phys. Lett.*, **137** (1987) 425.
[2] J.H.D. Eland, S.S.W. Ho and H.L. Worthington, *Chem. Phys.*, **290** (2003) 27.
[3] G. Dawber et al., *J. Phys. B: At. Mol. Opt. Phys.*, **27** (1994) 2191.
[4] D. Edvardsson et al., *Chem. Phys. Lett.*, **256** (2003) 341.
[5] M.J. Besnard et al., *J. Chem. Phys.*, **85** (1986) 1316.
[6] M.J. Besnard, L. Heller and G. Dujardin, *J. Chem. Phys.*, **89** (1988) 6554.
[7] D. Cossart, M. Bonneau and J.M. Robbe, *J. Mol. Spectrosc.*, **125** (1987) 413.
[8] L.G.M. Pettersson et al., *J. Chem. Phys.*, **96** (1992) 4884.

2.6 Oxygen, O_2

Molecular oxygen in its ground state has two unpaired electrons in the antibonding π_g orbital, so its electron configuration and term symbol are

$$1s\sigma_g{}^2 1s\sigma_u{}^2 2\sigma_g{}^2 2\sigma_u{}^2 3\sigma_g{}^2 1\pi_u{}^4 1\pi_g{}^2 \ldots\ldots\ldots\ldots\ldots\ldots ^3\Sigma_g{}^-.$$

The binding energies of the three outermost orbitals are about 12.3, 17 and 19 eV. Because of the unpaired electrons, many doubly ionised states can arise from removal of two electrons, including spin singlets, triplets and quintets. If this were not complication enough, the indirect dissociative double ionisation *via* superexcited states of atomic oxygen, as already seen in CO and NO, is not just strong, but dominant in photoionisation in the EUV (extreme ultraviolet) energy range accessed by HeII light. To show this and to illustrate a useful alternative form of coincidence map, in Fig. 2.6.1 we plot the sum of two coincident electron energies against their energy difference.

The diagonal lines in Fig. 2.6.1, which show atomic autoionisation, are stronger than even the most intense molecular double ionisation lines. Details of the processes involved are of great interest and have been studied in detail by several groups [1–4] but go beyond the scope of this volume. Fortunately, the electrons from atomic autoionisation

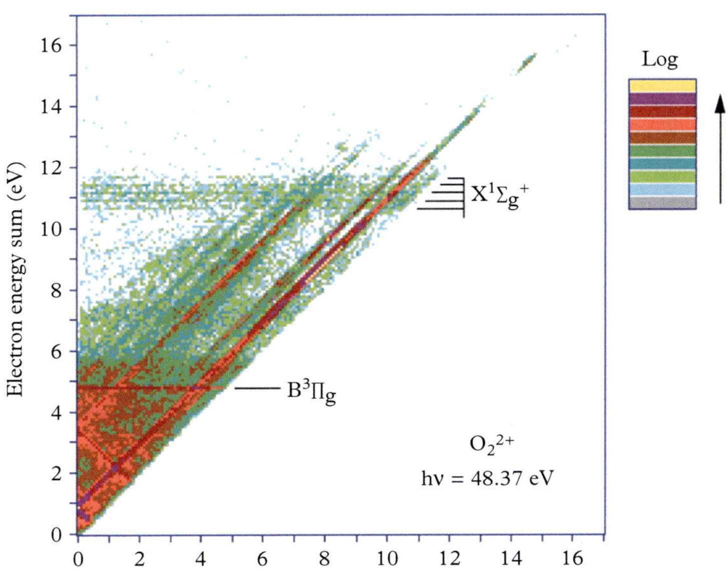

Figure 2.6.1 *Coincidence map for photoionisation of O_2 at 48.37 eV. In this form of map horizontal lines represent final doubly ionised states while all diagonal lines represent electrons of fixed energy coming from autoionisation. Two final $O_2{}^{2+}$ states are identified but others are obscured.*

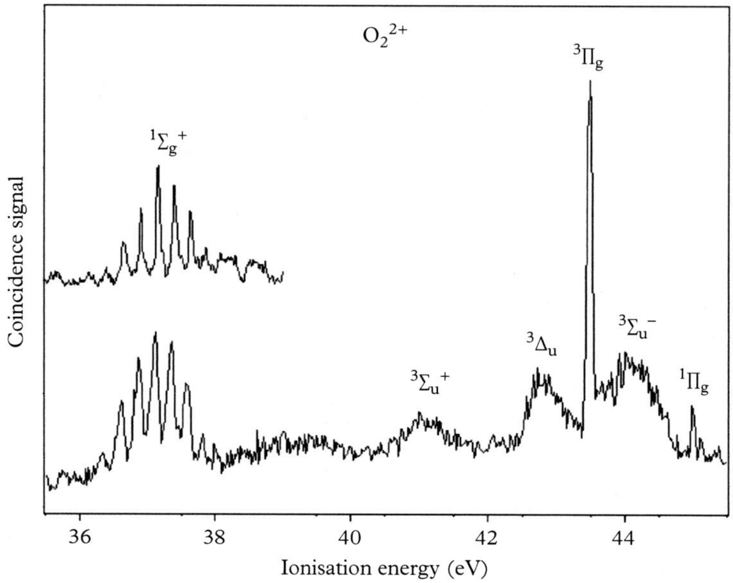

Figure 2.6.2 *TOF-PEPECO spectra of O_2 taken with HeIIα 40.81 and HeIIβ 48.37 eV light with elimination of most indirect double ionisation by excluding atomic autoionisation lines in the data analysis.*

come at fixed and sharp individual energies, which can be excluded in the data analysis. This exclusion may affect relative band intensities in the resulting spectrum of final states because more of the energy range for electron pairs is cut out from states at high binding energy than from those at lower ones. But any such effect within each individual band should be small since in direct double ionisation the pair distributions are generally smooth and flat. TOF-PEPECO spectra obtained by cutting out the indirect processes are shown as Fig. 2.6.2.

Essentially the same states were seen in the TPEsCO spectra of Hall et al. [5] and Dawber et al. [6], who carried out a vibrational analysis. The clearest view of most states of O_2^{2+} has come from the DFKER work of Lundqvist et al. [7], together with vital contributions from very extensive calculations of the molecular electronic structure. Simple estimates of ionisation energies from orbital energy and Coulomb repulsion energy are rather out of their depth in the case of O_2^{2+} except as a very rough guide. The X state, formed by removing the two antibonding electrons from the $1\pi_g$ orbital, has a shorter bond length than neutral oxygen and is well bound, so there is a long vibrational progression. Its centre, at about 37 eV implies a Coulomb repulsion energy of about 12 eV, when twice the binding energy of 12.3 eV in $1\pi_g$ is subtracted. This Coulomb energy agrees exactly with the value expected for two charges separated by the equilibrium bond length in neutral O_2 of 1.2 Å. The very prominent and apparently bound state of O_2^{2+} identified in Fig. 2.6.2 as $^3\Pi_g$ coming mainly from $3\sigma_g^{-1}1\pi_g^{-1}$ is at an ionisation energy of 43.55 eV, while addition of the same 12 eV Coulomb energy to two orbital energies, 12.3 and 19 eV, gives 41.3 eV. In contrast to the ground state, the first excited state of O_2^{2+}, $A^3\Sigma_u^+$, which may be expected on the same basis to lie near 41.3 eV, is entirely unbound, as shown also by Pettersson and Larsson [8], so can appear only as a continuous structure in the DFKER spectrum. Some resolved higher states of O_2^{2+} were identified clearly in the DFKER work [7], and others were seen in the Auger spectra measured and analysed by Larsson et al. [9] updating the interpretation of Sambe and Ramaker [10].

In the TOF-PEPECO spectrum taken at 80 eV photon energy, two of the higher states identified by Larsson et al. [9] are clearly visible. There is much less interference from atomic autoionisation at this photon energy, so the complete raw spectrum is shown in Fig. 2.6.3. Because of the lower energy resolution caused by the higher electron energies no vibrational structure is seen and the inherently broad and unresolved bands such as the one for the $A^3\Sigma_u^+$ state are strongly enhanced while the inherently very sharp peak for $B^3\Pi_g$ is diminished in apparent intensity. The band for the ground state, $X^1\Sigma_g^+$, has a distorted intensity profile, probably caused by Auger decay to this state from the inner valence states of singly charged O_2^+ at ionisation energies near 40 eV, which are populated strongly at this photon energy.

The relative intensities of bands in spectra from the different techniques cannot be compared directly because the techniques each involve different ionisation mechanisms and selection rules. The direct photoionisation in TOF-PEPECO populates singlet and triplet states when closed shell molecules are ionised, and in O_2 double ionisation it can populate singlet, triplet and quintet states including both Σ^+ and Σ^- states. In TPEsCO, double ionisation is almost entirely indirect and its selection rules are not known, while the inner shell Auger process is limited in the case of O_2 to triplet final states and excludes

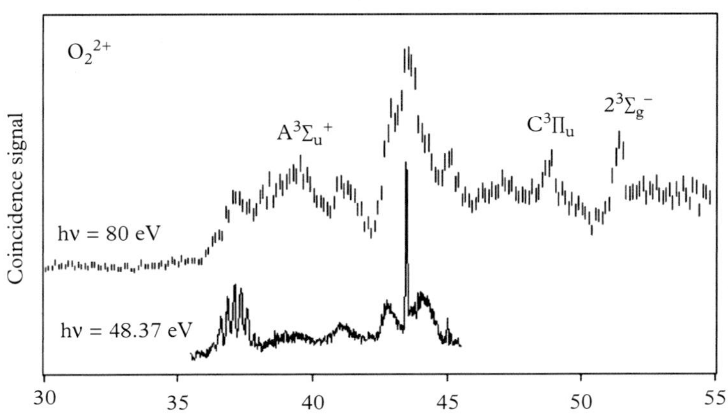

Figure 2.6.3 *The complete TOF-PEPECO spectrum of O_2 at 80 eV photon energy (upper spectrum), compared with the filtered spectrum at 48.37 eV from Fig. 2.6.2.*

states of Σ^+ symmetry [9, 10]. The electron impact ionisation used in the DFKER measurements can populate all final doubly charged states, but only quasi-bound states, which predissociate relatively slowly, can be seen clearly.

States of O_2^{2+} (the nomenclature is idiosyncratic but generally accepted)

State	DIE (eV) Adiabatic	Vertical	Ionisation
$X^1\Sigma_g^+$	(36.10)*	37.1	$1\pi_g^{-2}$
$A^3\Sigma_u^+$		39.5	$1\pi_u^{-1}1\pi_g^{-1}$
$W^3\Delta_u$	**42.6**	42.8	
$B^3\Pi_g$	**43.55**		$3\sigma_g^{-1}1\pi_g^{-1}$
$B^3\Sigma_u^-$		44.0	
$1^1\Pi_g$	**45.07**		$3\sigma_g^{-1}1\pi_g^{-1}$
$1^1\Delta_u$	**45.08**	45.25	
$C^3\Pi_u$	**48.90**		$2\sigma_u^{-1}1\pi_g^{-1}$
$2^3\Sigma_g^-$		51.5	$3\sigma_g^{-2}$
$1^5\Sigma_u^-$	**54.13**		

*According to Hall et al. [5] the adiabatic DIE is at 35.99 eV. In the TOF-PEPECO spectra the first reliably located peak is at 36.36 eV and the vibration spacing is 0.26 eV, so if the v=0 peak is one quantum lower, it must be at 36.1 eV. The most intense peak (vertical DIE) is at 37.1 eV.

References

[1] P. Bolognesi et al., *Phys. Rev. Lett.*, **82** (1999) 2075.
[2] R. Feifel and J.H.D. Eland, *Phys. Rev. A*, **71** (2005) 034501.
[3] R. Feifel, J.H.D. Eland and D. Edvardsson, *J. Chem. Phys.*, **122** (2005) 144308.
[4] A.S. Sandhu et al., *Science*, **322** (2008) 1081.
[5] R. Hall et al., *Phys. Rev. Lett.*, **68** (1992) 2751.
[6] G. Dawber et al., *J. Phys. B: At. Mol. Opt. Phys.*, **27** (1994) 2191.
[7] M. Lundqvist et al., *J. Phys. B: At. Mol. Opt. Phys.*, **29** (1996) 499.
[8] L.G.M. Pettersson and M. Larsson, *J. Chem. Phys.*, **94** (1991) 818.
[9] M. Larsson et al., *J. Phys. B: At. Mol. Opt. Phys.*, **23** (1990) 1175.
[10] H. Sambe and D.E. Ramaker, *Chem. Phys.*, **104** (1986) 331.

2.7 Bromine, Br_2

Molecular bromine has four electrons filling the antibonding π_g orbital, so its ground state is a singlet with electron configuration in the outer shells and term symbol as

$$\sigma_u{}^2\sigma_g{}^2\pi_u{}^4\pi_g{}^4 \ldots\ldots\ldots\ldots\ldots\ldots {}^1\Sigma_g{}^+.$$

When two electrons are removed from a full π-orbital of a linear molecule we always expect to generate three electronic states with term symbols ${}^3\Sigma^-$, ${}^1\Delta$ and ${}^1\Sigma^+$ in the familiar LS coupling notation (all *gerade* in a centrosymmetric molecule). These are expected to be the lowest energy states in $Br_2{}^{2+}$. But for accurate theoretical calculation of the energy levels in a molecule with atoms as heavy as Br this scheme is not entirely adequate, as relativistic corrections are necessary. When spin–orbit coupling is strong the spin and orbital angular momenta are no longer good quantum numbers, and the Σ, Π, ${}^1\Delta$... symbols should strictly be discarded in favour of total angular momentum and parity. Nevertheless we persist in using the familiar designations for the TOF-PEPECO spectra of $Br_2{}^{2+}$ in Fig. 2.7.1.

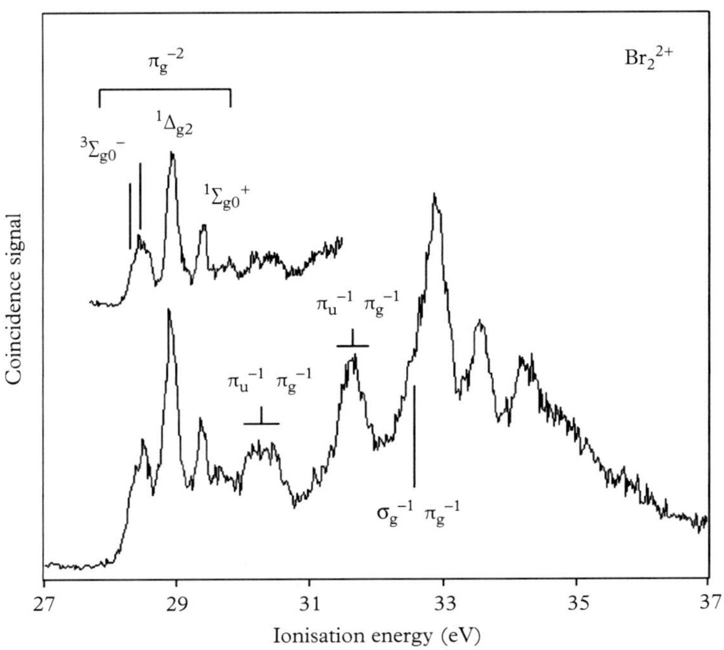

Figure 2.7.1 *TOF-PEPECO spectra of Br_2 taken with HeIIα 40.81 eV (lower trace) and NeIII 32.69 eV light. The bands where no orbital ionisation is given are attributed to multiple states with mixed configurations.*

The three lowest energy bands were first reported in a TPEsCO study [1] which was followed by TOF-PEPECO work which included spectra from photoionisation at three wavelengths, accompanied by relativistic calculations [2]. Despite the good energy resolution of measurements with the lowest energy photons, no vibrational structure could be discerned; this was attributed to the small calculated vibrational spacing of 34 meV in addition to electronic state congestion. For this reason the experimental double ionisation energies (DIEs) are listed below as vertical values, though in some cases the adiabatic values may be the same. A later set of calculations [3] gave equally good agreement on the electronic state energies.

States of Br_2^{2+} (the usual state designations are given first, then relativistic ones)

State(s)		DIE (eV)	Orbitals ionised
$X^3\Sigma_{g0}^-$	0_g^+	28.39	π_g^{-2}
$^3\Sigma_{g1}^-$	1_g	28.53	π_g^{-2}
$^1\Delta_{g2}$	2_g	28.91	π_g^{-2}
$^1\Sigma_{g0}^+$	0_g, 0_u, 3_u	29.38	π_g^{-2}, $\pi_u^{-1}\pi_g^{-1}$
$^3\Sigma_u^+$	1_u, 0_u	31.6	$\pi_u^{-1}\pi_g^{-1}$
Multiple		32.6	$\sigma_g^{-1}\pi_g^{-1}$
Multiple		32.9	
Multiple		33.6	
Multiple		34.2	

References

[1] A.J. Yencha, S.P. Lee and G.C. King, in Proceedings of the XXIV ICPEAC Conference, 2005.

[2] T. Fleig et al., *Chem. Phys.*, **343** (2008) 270.

[3] L. Streit, F.B.C. Marchado and R. Custodio, *Chem. Phys. Lett.*, **506** (2011) 22.

2.8 Iodine monochloride, ICl

Iodine monochloride has the same configuration in its outer shells as bromine, but the distinction between gerade and ungerade orbital symmetry disappears, so we identify the outermost π-orbital with an asterisk:

$$\sigma^2\pi^4\pi^{*4}\ \ldots\ldots\ldots\ldots\ldots\ldots\ {}^1\Sigma_g{}^+.$$

Vertical single ionisation energies from these orbitals are known from the photoelectron spectrum [1] and threshold photoelectron spectrum [2] as about 10.2, 12.8 and 14.3 eV respectively. An experimental problem is that the compound is always in dynamic equilibrium with its constituent elements:

$$2\mathrm{ICl} \leftrightarrow \mathrm{I_2} + \mathrm{Cl_2}.$$

The vapour pressure of ICl is greater than that of iodine, but the volatility of chlorine at room temperature is much greater than that of ICl. The seriousness of this problem was unfortunately overlooked when the TOF-PEPECO spectrum of ICl was first measured [3], and the origin of some bands at energies above 30 eV was not investigated fully. We can now see that as their positions coincide exactly with the principal low-energy bands in the TPEsCO spectrum of molecular chlorine [4], they are probably not attributable to ICl. We therefore show only the part of the spectrum below 31 eV in Fig. 2.8.1.

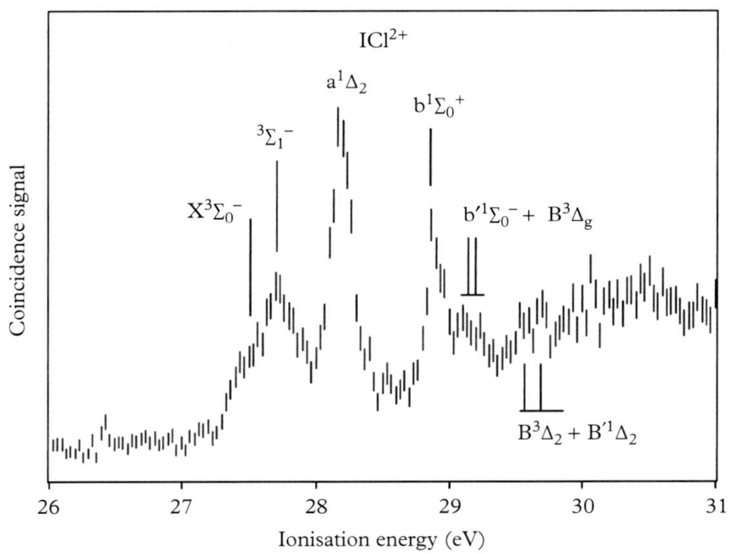

Figure 2.8.1 *TOF-PEPECO spectra of ICl taken with NeIII 32.69 eV light. In this figure, the coincidence signals are represented as error bars showing the statistical uncertainty of ± 1σ at each point.*

The spectrum taken with NeIII (32.69 eV) light is shown using error bars for the coincidence signal, because the number of coincidence counts at each point is relatively low. This is because this NeIII line is less intense in the laboratory lamps than the more usual HeII lines. The purpose of using it is to improve the electron energy resolution by lowering the energy, for which it is particularly suitable because of being a singlet. There are more intense lines of NeII at low photon energy, but they are doublets because of spin–orbit coupling and cannot be separated without greater loss of intensity.

To interpret the spectrum, relativistic calculations with extensive electron correlation have been carried out [3], leading to the state identifications shown in the figure and the list below. All the identified lines arise from ionisation out of the π-orbitals. In single ionisation [2] the bands from both the π- and π^*-orbitals have vibrational structure showing that electrons in these orbitals are weakly rather than strongly bonding, whereas ionisation from the σ-orbital shows no structure. We can therefore surmise that doubly ionised states involving σ ionisation are completely unbound and will contribute only broad unstructured bands to the double ionisation spectrum.

States of ICl^{2+}

State(s)	DIE (eV)	Orbitals ionised
$X^3\Sigma_0^-$	27.45	π^{*-2}
$X^3\Sigma_1^-$	27.69	π^{*-2}
$a^1\Delta_2$	28.14	π^{*-2}
$b^1\Sigma_0^+$	28.84	π^{*-2}
$b'^1\Sigma_0^-$ and $B^3\Delta_3$	29.07	$\pi^{-1}\pi^{*-1}$
$B^3\Delta_2$ and $^1\Delta_2$	29.62	$\pi^{-1}\pi^{*-1}$

References

[1] S. Evans and A.F. Orchard, *Inorg. Chim. Acta*, 5 (1971) 81.
[2] A.J. Yencha, M.C.A. Lopes and G.C. King, *Chem. Phys. Lett.*, **325** (2000) 559.
[3] D. Edvardsson et al., *Chem. Phys.*, **332** (2007) 249.
[4] A.G. McConkey et al., *J. Phys. B: At. Mol. Opt. Phys.*, **27** (1994) 271.

2.9 Iodine, I_2

Molecular iodine has so many electrons overall that we mention only the outermost shells to describe its electron configuration and term symbol:

$$4p^{12}4d^{20}\sigma_u^2\sigma_g^2\pi_u^4\pi_g^4 \ldots\ldots\ldots\ldots\ldots\ldots {}^1\Sigma_g^+.$$

The 4d shell has a binding energy of about 60 eV, so it is near to becoming involved in valence interactions, while 4p is at about 130 eV, as can be seen in a wide-range photoelectron spectrum of iodine [1].

The TOF-PEPECO spectrum has been measured at three photon energies of which two are shown in Fig. 2.9.1, and it has been compared with relativistic calculations on the low-energy states [2]. The first four peaks all arise from π_g^{-2} ionisation and of the states expected from $\pi_u^{-1}\pi_g^{-1}$ ionisation we can identify one clear peak, ${}^3\Delta_u$, and a shoulder near 26 eV attributed in the table below to two states. As in identifying the states for Br_2, we retain the familiar state designations for the bands that have been identified. The calculations indicate that there are many overlapping states of doubly charged iodine at higher energies. Using approximate orbital energies from the valence photoelectron spectrum [3] we can estimate that states from $\pi_u^{-1}\pi_g^{-1}$ ionisation will lie mainly

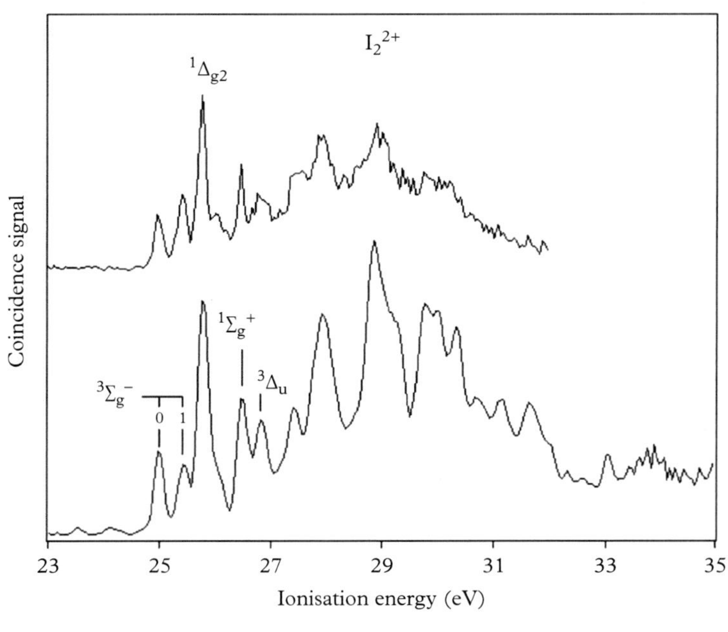

Figure 2.9.1 *TOF-PEPECO spectra of I_2 taken with HeIIα 40.81 eV (lower trace) and NeIII 32.69 eV light (upper trace). The bands where no state designation is given are attributed to multiple states with mixed configurations.*

in the range 26–29 eV, those from π_u^{-2} and $\sigma_g^{-1}\pi_g^{-1}$ ionisation in the range 27–30 eV, those from $\sigma_g^{-1}\pi_u^{-1}$ above 30 eV and the single $^1\Sigma_g^+$ state from σ_g^{-2} ionisation around 33.5 eV. It is reasonable that the intense bands occur in the range where the degenerate π-orbitals can contribute, but it would be dangerous to attempt any closer attribution without detailed calculations. The orbital picture may be a poor approximation for the high energy states of this and many dications, because accurate description of their electronic structure requires inclusion not only of many two-hole configurations like those suggested above, but also of numerous three-hole one-particle configurations including an electron in a normally unoccupied orbital.

States of I_2^{2+}

State(s)	**DIE (eV)**	**Orbitals ionised**
$X^3\Sigma_{g0}^-$	24.95	πg^{-2}
$^3\Sigma_{g1}^-$	25.4	πg^{-2}
$^1\Delta_{g2}$	25.74	πg^{-2}
$^1\Sigma_{u0}^-$ and $^3\Sigma_{u1}^+$	26.02	$\pi u^{-1}\pi g^{-1}$
$^1\Sigma_{g0}^+$	26.3	πg^{-2}
$^3\Delta_u$	26.82	$\pi u^{-1}\pi g^{-1}$
Unidentified	27.4, 28.0, 28.9, 29.9, 30.3, 30.7, 31.1, 31.7, 33.1	

References

[1] S. Aksela, H. Aksela and T.D. Thomas, *Phys. Rev. A*, **19** (1979) 721.
[2] D. Edvardsson et al., *Chem. Phys.*, **324** (2006) 674.
[3] S. Evans and A.F. Orchard, *Inorg. Chim. Acta*, **5** (1971) 81.

Chapter 3

Triatomic molecules

3

Triatomic molecules

With three atoms, the triatomic molecules present an added degree of complexity. Besides potentially having many more electrons they all have three vibrational degrees of freedom (three normal modes) instead of the single one in a diatomic molecule. In vertical transitions from the ground state, excitation should be confined to symmetric modes in the absence of vibronic coupling, but this restriction is strong in practice only for linear ABA molecules where the symmetric stretch alone should be excitable in single-step transitions. For asymmetric and bent triatomic molecules multiple modes can be excited, so the spectral bands may be congested in all forms of electronic spectra, including double ionisation. With the resolution currently available in TOF-PEPECO, we can therefore expect to see vibrationally resolved bands mainly in spectra of the linear molecules.

The same processes of indirect double ionisation that are seen in spectra of the diatomic molecules operate just as much for these larger molecules. They are more difficult to distinguish from direct molecular double ionisation when the observed bands are broad bumps without vibrational structure, so for some of these molecules it is impossible to determine meaningful adiabatic double ionisation energies.

The order of the molecules in this chapter is set first by the number of valence electrons, then by the molecular weight.

Double Photoionisation Spectra of Molecules. John H. D. Eland and Raimund Feifel.
© John H. D. Eland and Raimund Feifel, 2018. Published in 2018 by Oxford University Press.
DOI 10.1093/oso/9780198788980.001.0001

3.1 Water, H_2O

The water molecule has two equivalent O—H bonds with an angle of 104° between them. Its electron configuration in C_{2v} symmetry is

$$O1s^2 2a_1{}^2 1b_2{}^2 3a_1{}^2 1b_1{}^2 \ldots\ldots\ldots\ldots\ldots\ldots\ldots {}^1A_1.$$

Here the $2a_1$ orbital is mainly based on O2s, and is of inner valence character, and the three outer valence orbitals are all non-degenerate and doubly occupied. Because of water's exceptional importance, its double ionisation had been studied by several techniques including Auger spectroscopy [1], ion-ion coincidence spectroscopy [2, 3] and double charge transfer (DCT) [2, 4] before the advent of TOF-PEPECO. The ion–ion coincidence measurements [2] already suggested that indirect dissociative processes must be in operation, as double photoionisation is observed at energies far below the molecular double ionisation energy (DIE). Because no stable or metastable H_2O^{2+} ions have ever

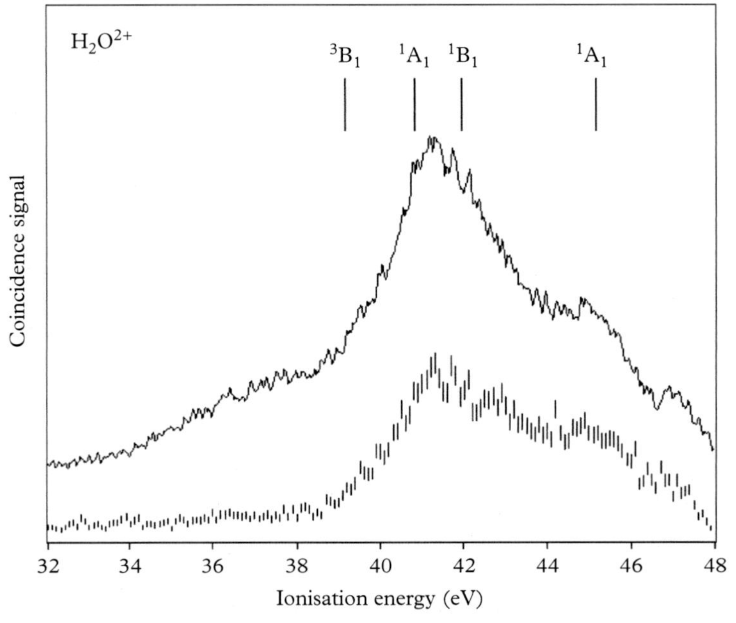

Figure 3.1.1 *TOF-PEPECO spectra of water at 48.37 eV photon energy, taking all electrons (upper trace) or including only pairs with an energy difference less than 2 eV (lower trace). The lower trace is thought to show mainly direct (molecular) double photoionisation. Calculated positions of some states of H_2O^{2+} are indicated; ${}^3B_1$ and ${}^1B_1$ are from $3a_1{}^{-1}1b_1{}^{-1}$ ionisation while the ${}^1A_1$ states are from $1b_1{}^{-2}$ and $3a_1{}^{-2}$. Error bars (2σ) are used to represent the reduced numbers of coincidence counts in the lower trace.*

been observed, the molecular DIE is a slippery concept as discussed in the Introduction, but it is reliably calculated to be at about 39 eV. That indirect double ionisation is dominant is demonstrated by TOF-PEPECO measurements [5] and is illustrated by the spectra in Fig. 3.1.1.

The TOF-PEPECO spectrum shown in the upper trace of Fig. 3.1.1 was obtained by taking all electron pairs extends almost down to the thermodynamic limit of 31.7 eV for formation of H^+ + OH^+ pairs with no internal or kinetic energy, so an indirect dissociative process must be involved. The coincidence map also shows strong indirect double ionisation, including the presence of fixed-energy electrons characteristic of superexcited O*, just as in the double photoionisation of CO, O_2 and NO. To reduce the contribution of these indirect processes to the spectrum, electron pairs with a large energy difference between them have been excluded in deriving the lower trace in Fig. 3.1.1. This strategy is effective because the majority of indirect and dissociative double ionisation arises from singly charged H_2O^+ states lying within the energy range of true molecular double ionisation. These states are concentrated at the lower energy end of this range and show up in the electron pair distributions as an excess of intensity for near-zero energies of one electron and near-maximum energy for the other. When low-energy electrons

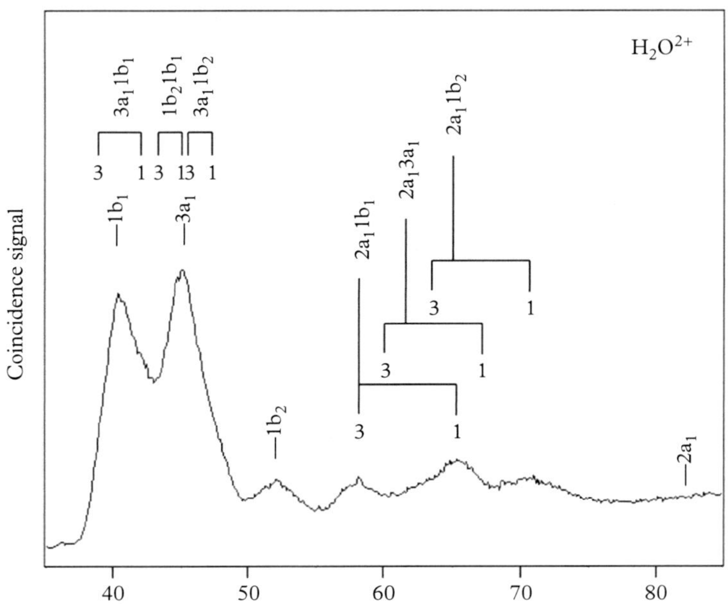

Figure 3.1.2 *TOF-PEPECO spectrum of water at 110 eV photon energy. Bars above the spectrum show the calculated energies [10] of doubly ionised states from ionisation out of the different orbitals. Where only a single orbital is identified, both electrons came from that same orbital producing a singlet state.*

are excluded the contribution of indirect double ionisations to the spectrum is strongly reduced, while direct double ionisation, because of its more even and continuous distribution of electron energies, is hardly affected. Molecular structure calculations including electron correlation effects [6, 7] confirm that there are no electronic states of the dication with potential minima capable of supporting bound levels, so the states with positions suggested in Fig. 3.1.1 contribute only unstructured bands to the spectra.

For a view of the double photoionisation of water over a wider range, Fig. 3.1.2 shows the spectrum taken with synchrotron light at a photon energy of 110 eV. This spectrum bears a strong resemblance to the Auger spectrum [1], which has been interpreted several times with the help of molecular structure calculations including correlation effects [8–10]. The different calculations agree closely on the energies of the doubly ionised states, and include triplets as well as singlets. As the figure demonstrates, the calculated ionisation energies match the experimental spectrum well and show that many of the bands overlap. Indirect double ionisation, so prominent at 48.37 eV, is less noticeable in the spectrum at this higher energy, but still present and visible in the coincidence maps. At even higher energy, Hikosaka et al. [11] have shown that indirect double ionisation of water involving O^* occurs again by a different mechanism, namely double ionisation of water by the Auger effect from states with a vacancy in the O1s orbital [8].

References

[1] W.A. Moddeman et al., *J. Chem. Phys.*, **55** (1971) 2317.

[2] P.J. Richardson et al., *J. Chem. Phys.*, **84** (1986) 3189.

[3] D. Winkoun et al., *J. Phys. B: At. Mol. Opt. Phys.*, **21** (1988) 1385.

[4] F. Alvarado, R. Hoekstra and T. Schlathölter, *J. Phys. B: At. Mol. Opt. Phys.*, **38** (2005) 4085.

[5] J.H.D. Eland, *Chem. Phys.*, **323** (2006) 391.

[6] D.L. Cooper et al., *J. Chem. Phys.*, **87** (1987) 1666.

[7] B. Gervais et al., *J. Chem. Phys.*, **131** (2009) 024302.

[8] H. Siegbahn, L. Asplund and P. Kelve, *Chem. Phys. Lett.*, **35** (1975) 330.

[9] V. Carravetta and H. Ågren, *Phys. Rev. A*, **35** (1987) 1022.

[10] L. Inhester et al., *J. Chem. Phys.*, **136** (2012) 144304.

[11] Y. Hikosaka et al., *J. Chem. Phys.*, **137** (2012) 191101.

3.2 Hydrogen sulphide, H_2S

Although the valence electron configuration of H_2S is the same as that of water, its bond angle is only 92°. The configuration, omitting the inner shells on the S atom, is

$$4a_1{}^2 2b_2{}^2 5a_1{}^2 2b_1{}^2 \ldots\ldots\ldots\ldots\ldots\ldots\ldots {}^1A_1.$$

Here double ionisation from the outermost orbital produces a bound state of the dication with well-resolved vibrational structure. This has been known for many years from the Auger spectrum [1], where the Auger decay and subsequent dissociation have since been studied in considerable detail [2–6]. The resolved vibrational levels were seen using extreme ultraviolet (EUV) light for photoionisation in a TPEsCO experiment [7] and then more clearly by TOF-PEPECO [8, 9].

A combined TPEsCO and ion–ion coincidence study [7] already showed that indirect double ionisation is prevalent in H_2S, because ion pairs and threshold electron pairs both appear at photon energies well below the molecular double ionisation energy. Coincidence maps from TOF-PEPECO [8, 9] show that a large part of this is autoionisation

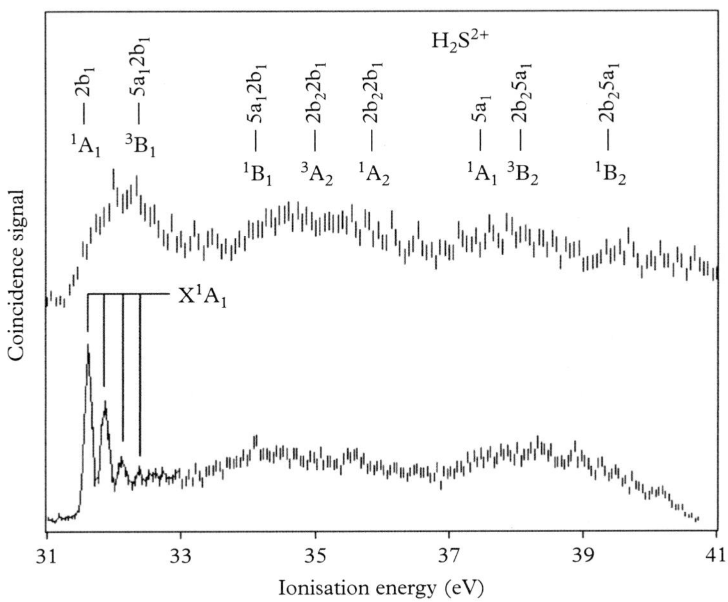

Figure 3.2.1 *TOF-PEPECO spectra of H_2S at 40.81 eV (lower trace) and 100 eV (upper). Because of the low number of counts after exclusion of indirect double ionisation, data are plotted as error bars, with a line to guide the eye through the resolved structure. The apparent peak at 34.1 eV is an artefact. Theoretical state energies and configurations [1], shifted to match the first peak, are shown above the spectra as in Fig. 3.1.1.*

of superexcited atomic sulphur. In Fig. 3.2.1 this autoionisation has been largely suppressed by selecting electron pairs where the electron energy difference is limited to a small proportion of the total energy. Calculated H_2S^{2+} state energies [1] and combination of orbital energies from the photoelectron spectrum [10] with a Coulomb energy show that there must be seven states within the range covered by the figure. All but the lowest state are evidently repulsive in the Franck–Condon zone and give only broad overlapping bands. There is some correspondence between the calculated state energies and broad bands of intensity in the spectrum, but no definite assignments can be made. A similar correspondence was found between the theoretical and experimental Auger spectra from photoionisation to specific spin-orbit and molecular field components of the $S2p^{-1}$ hole states [9].

Ground state of H_2S^{2+}

State	DIE (eV)
X^1A_1	31.66

References

[1] A. Cesar et al., *J. Chem. Phys.*, **93** (1990) 918.
[2] S. Svensson et al., *Phys. Rev. Lett.*, 72 (1994) 3021.
[3] F. Gel'mukhanov et al., *Phys. Rev. A*, **53** (1996) 1379.
[4] A.M. Bueno et al., *Phys. Rev. A*, **67** (2003) 022714.
[5] R. Guillemin et al., *J. Chem. Phys.*, **122** (2005) 094318.
[6] M. Poygin et al., *Phys. Rev. A*, **74** (2006) 012711.
[7] J.H.D. Eland et al., *J. Phys. B: At. Mol. Opt. Phys.*, **30** (1997) 2177.
[8] J.H.D. Eland, *Chem. Phys.*, **294** (2003) 171.
[9] J.H.D. Eland et al., *Phys. Chem. Chem. Phys.*, **13** (2011) 18428.
[10] P. Baltzer et al., *Chem. Phys.*, **195** (1995) 403.

3.3 Hydrogen cyanide, HCN

HCN is a linear molecule, but its photoelectron spectrum is notoriously difficult to analyse [1, 2] and unfortunately the same is true of its double ionisation spectrum measured by TOF-PEPECO. Although its valence electron configuration

$$N1s^2C1s^23\sigma^24\sigma^21\pi^45\sigma^2 \ldots\ldots\ldots\ldots\ldots\ldots {}^1\Sigma^+$$

is nominally the same as those of N_2 and CO, the two outermost orbitals are very close together in energy and strong vibronic interactions seriously complicate the single ionisation spectrum. An even less amenable proximity of several states on the energy scale is found, as expected, in double ionisation. Removal of two electrons from 5σ will produce a ${}^1\Sigma^+$ state, removal of one each from 1π and 5σ will produce ${}^3\Pi$ and ${}^1\Pi$ and removal of two from 1π will give ${}^3\Sigma^-$, ${}^1\Delta$ and another ${}^1\Sigma^+$. These six states might all be found in the low-energy part of the spectrum below 40 eV, but detailed calculations [3] indicate that only one of the ${}^1\Sigma^+$ states is likely to be seen. Because the orbitals, particularly 1π, are involved in bonding mainly between C and N, substantial lengthening of this bond is expected in double ionisation, with consequent long vibrational progressions in the C—N vibration and possibly low intensity for the adiabatic 0–0 transitions.

In Fig. 3.3.1 the spacings between peaks in what look like vibrational progressions are about 125 meV ($\sim$ 1,000 cm^{-1}), substantially reduced, as expected, from the C—N

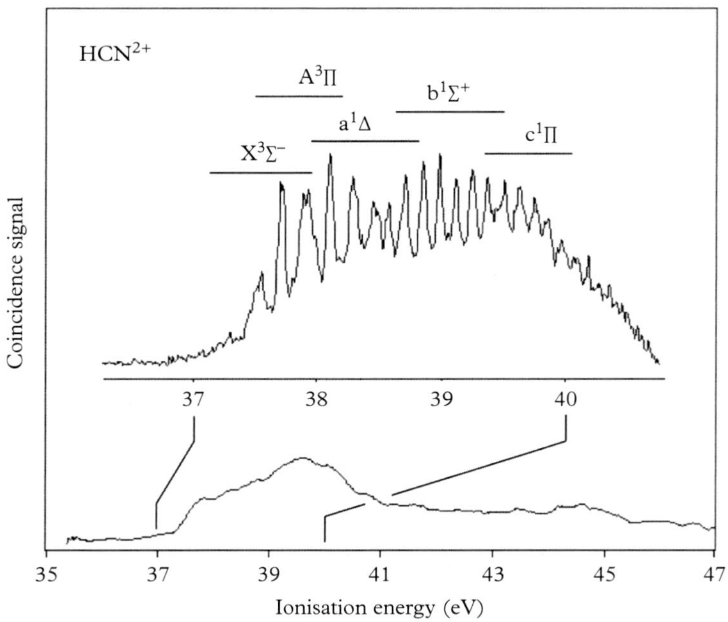

Figure 3.3.1 *TOF-PEPECO spectra of HCN at 48.37 eV (lower) and 40.81 eV (upper and expanded inset). Above the spectrum the approximate ranges where different states contribute are indicated.*

stretching frequency in neutral HCN of 259 meV. The spectrum of DCN has also been measured at 40.81 eV but looks almost the same as that of HCN, confirming that the C—N and not C—H vibration is dominantly excited. Despite high-level electronic and geometric molecular structure calculations [3], we cannot absolutely pin down which lines belong to which of the electronic states, nor give definite values of adiabatic double ionisation energies. But the calculations predict very clearly that the energy order of the states, each at its own optimum geometry, is $X^3\Sigma^-$, $A^3\Pi$, $a^1\Delta$, $b^1\Sigma^+$ and then $c^1\Pi$. At the geometry of neutral HCN (the middle of the Franck–Condon zone) the order is essentially the same, but $X^3\Sigma^-$ and $A^3\Pi$ are very close together. The first distinct peak, near 37.6 eV, is clearly composite in the spectra of both HCN and DCN and so must have contributions from $X^3\Sigma^-$ and $A^3\Pi$. The structured part of the spectrum, after subtraction of a smooth background contribution, can be simulated by a summation of five vibrational progressions with different vibration intervals, but similar Franck–Condon envelopes. Unfortunately, more than one such simulation can be made to fit the observed spectra equally well. On the basis of these observations, together with the theoretical results, we can define the energy ranges in which we believe that each state plays its part, and these are marked out in the figure. The coincidence maps in this case show almost no contribution from indirect processes of dissociation and autoionisation. Much better energy resolution will be needed to fully disentangle the double ionisation spectrum of HCN and for that, future investigators may need an entirely new technique.

References

[1] H. Köppel et al., *Chem. Phys.*, **37** (1979) 303.
[2] J.H.D. Eland et al., *Chem. Phys.*, **229** (1998) 149.
[3] M. Hochlaf, A. Pilcher-Clayton and J.H.D. Eland, *Chem. Phys.*, **309** (2005) 291.

3.4 Carbon dioxide, CO_2

With three heavy atoms, carbon dioxide has a richer electronic structure than lighter molecules. Its electron configuration is

$$O1s^4C1s^23\sigma_g{}^22\sigma_u{}^24\sigma_g{}^23\sigma_u{}^21\pi_u{}^41\pi_g{}^4 \ldots\ldots\ldots\ldots\ldots\ldots\ldots {}^1\Sigma_g{}^+.$$

The binding energies of the four outermost orbitals are near 14, 17.5, 18 and 19 eV, according to the photoelectron spectrum [1]. The outermost π-orbital is non-bonding and located purely on the oxygen atoms, whereas the inner π-orbital is the most strongly bonding of all the valence orbitals. From combination of the orbitals we can expect about 30 doubly charged states of $CO_2{}^{2+}$ at energies below 50 eV. The energies of these states have been calculated several times, with reasonably concordant results [2–4]. The double ionisation spectrum of CO_2 is indeed rich in structure with multiple bands, though many states formed by ionisation of bonding electrons are undoubtedly missed because they form broad bands indistinguishable from an underlying background.

In Fig. 3.4.1, all three states from $1\pi_g{}^{-2}$ ionisation are clearly seen and have resolved vibrational structure, but of the six states expected from $1\pi_u{}^{-1}1\pi_g{}^{-1}$ ionisation, only

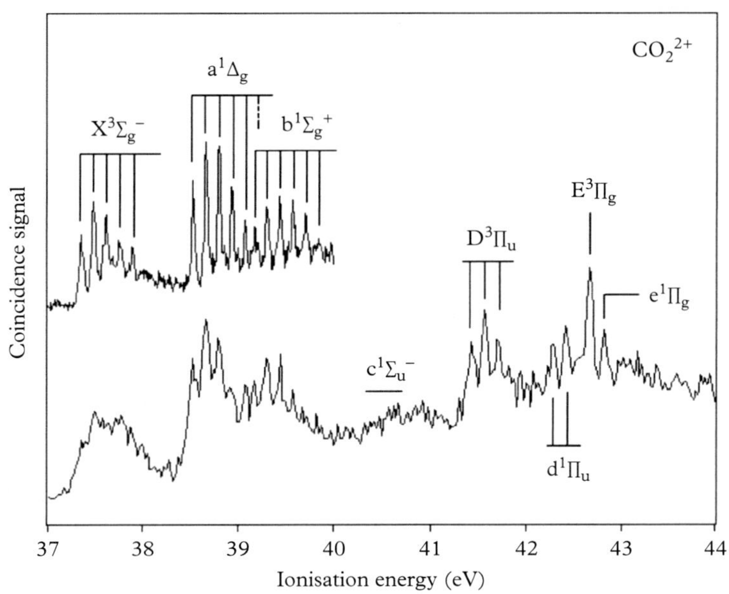

Figure 3.4.1 *TOF-PEPECO spectra of CO_2 at 40.81 eV (upper trace) and 48.37 eV (lower trace). Eight states are identified, but several more certainly contribute to the rising background. Two sharp but weak peaks were also seen near 46 eV but have not been identified.*

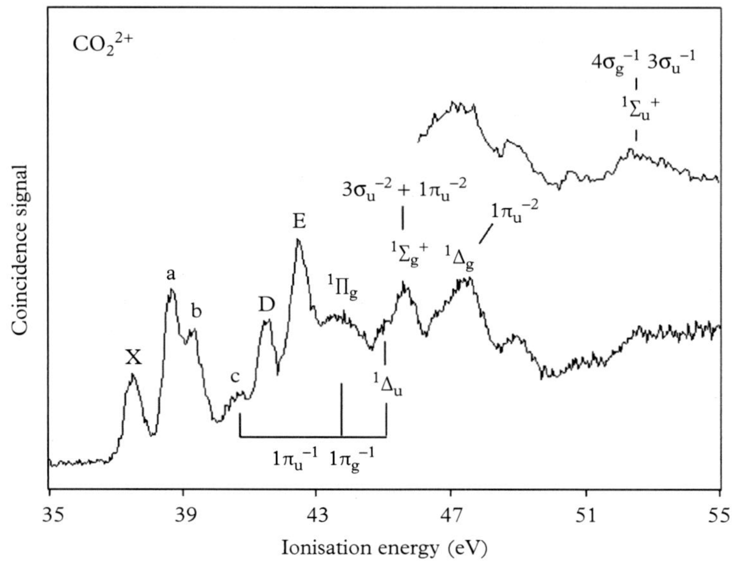

Figure 3.4.2 *TOF-PEPECO spectrum of CO_2 at 65 eV photon energy (main trace) and a section of a spectrum measured at 85 eV emphasising the state near 52 eV. For the states already identified in Fig. 3.4.1, only the names are given, while for higher states no name letters are given because other states of the same symmetry may intervene.*

$c^1\Sigma_u^-$, the lowest energy state from this configuration, is marked, but the others undoubtedly contribute in roughly the same energy range. There is no visible trace of the three states from $1\pi_u^{-2}$ ionisation, where both electrons come from the bonding orbital, although they are surely present also. The higher states giving resolved peaks and identified in Fig. 3.4.1 are formed by ionisation from the non-bonding π_g-orbital and one of the two weakly bonding σ-orbitals.

For wider coverage of the states of CO_2^{2+}, Fig. 3.4.2 shows a spectrum taken with 65 eV photons and a section measured with 85 eV photons. Because of the higher electron energies and the shorter flight path used at the synchrotron light source, resolution is poor by comparison with the previous figure, but more states can be seen.

At the higher photon energies and lower resolution the broad bands due to unbound electronic states become more prominent, and three from the configuration $1\pi_u^3 1\pi_g^3$ can be identified with help from numerical calculations. The higher energy states tend to have mixed configurations, and the identifications with bands in the spectrum are provisional. For the band at 49 eV, the calculations [2–4] suggest several possible identifications; it may be a blend of states.

The three low-energy states of CO_2^{2+} have been examined experimentally by TPEsCO [5, 6] and double charge transfer (DCT) [2, 7] as well as by TOF-PEPECO [6], and their vibronic structure has been analysed in detail [4]. Higher energy states are seen in the Auger spectrum [8, 9] and in wide-range TOF-PEPECO [10]. Doubly ionised states of the molecule can also be formed by Auger decay from states with carbon and oxygen 1s core holes, and some aspects of the subsequent dissociations have been studied in detail [11–13].

States of CO_2^{2+} (with adiabatic DIEs as bold)

State(s)	DIE (eV)	Orbitals ionised
$X^3\Sigma_g^-$	**37.34**	$1\pi_g^{-2}$
$a^1\Delta_g$	**38.52**	$1\pi_g^{-2}$
$b^1\Sigma_g^+$	**39.30**	$1\pi_g^{-2}$
$c^1\Sigma_u^-$	40.1	$1\pi_u^{-1}1\pi_g^{-1}$
$D^3\Pi_u$	**41.43**	$3\sigma_u^{-1}1\pi_g^{-1}$
$d^1\Pi_u$	**42.30**	$3\sigma_u^{-1}\pi_g^{-1}$
$E^3\Pi_g$	**42.65**	$4\sigma_g^{-1}1\pi_g^{-1}$
$e^1\Pi_g$	**42.82**	$4\sigma_g^{-1}1\pi_g^{-1}$
$^1\Pi_g$	43.7	$1\pi_u^{-1}1\pi_g^{-1}$
$^1\Delta_u$	45.1	$1\pi_u^{-1}1\pi_g^{-1}$
$^1\Sigma_g^+$	45.7	$3\sigma_u^{-2} + 1\pi_u^{-2}$
$^1\Delta_g$	47.4	$1\pi_u^{-2}$
??	49.0	
$^1\Sigma_u^+$	52.3	$4\sigma_g^{-1}3\sigma_u^{-1}$
??	62	

References

[1] P. Roy et al., *J. Chem. Phys.*, **84** (1986) 2050.
[2] P. Millié et al., *J. Chem. Phys.*, **84** (1986) 1259.
[3] H. Hogreve, *J. Phys. B: At. Mol. Opt. Phys.*, **28** (1995) L263.
[4] M. Hochlaf et al., *J. Phys. B: At. Mol. Opt. Phys.*, **31** (1998) 2163.
[5] R.I. Hall et al., *Chem. Phys.*, **187** (1994) 125.
[6] A.E. Slattery et al., *J. Chem. Phys.*, **122** (2005) 084317.
[7] L. Langford et al., *Chem. Phys.*, **149** (1991) 445.
[8] W.E. Moddeman et al., *J. Chem. Phys.*, **55** (1971) 2317.
[9] R. Püttner et al., *J. Phys. B: At. Mol. Opt. Phys.*, **41** (2008) 045103.
[10] J.H.D. Eland et al., *Chem. Phys. Lett.*, **646** (2016) 31.
[11] G. Öhrwall et al., *J. Phys. B: At. Mol. Opt. Phys.*, **35** (2002) 4543.
[12] Z.D. Pešić et al., *Phys. Rev. A*, **78** (2008) 051401.
[13] J. Laksman et al., *J. Chem. Phys.*, **136** (2012) 104303.

3.5 Nitrous oxide, N_2O

Nitrous oxide has the linear structure NNO and its electron configuration is

$$\mathrm{O1s^2N1s^44\sigma^25\sigma^26\sigma^21\pi^47\sigma^22\pi^4} \ldots\ldots\ldots\ldots\ldots\ldots\ldots\ {}^1\Sigma^+.$$

The binding energies of the four outermost orbitals are about 12.8, 16.2, 18 and 20 eV, according to the photoelectron spectrum [1]. The outermost π-orbital and the two σ-orbitals appear in the spectrum as non-bonding while the inner π-orbital is the most strongly bonding of the valence orbitals. From combinations of these orbital energies we can expect five doubly ionised states at binding energies below 40 eV, and detailed calculations [2] bear this out. Nitrous oxide is a convenient experimental and theoretical subject and has been studied extensively in both its singly and doubly ionised forms.

The spectra displayed in Fig. 3.5.1 demonstrate that few of the states of N_2O^{2+} are stable or metastable, or have deep enough potential wells to support vibrational levels. Extensive high-level calculations by Levasseur and Millié [3] and by Taylor et al. [2]

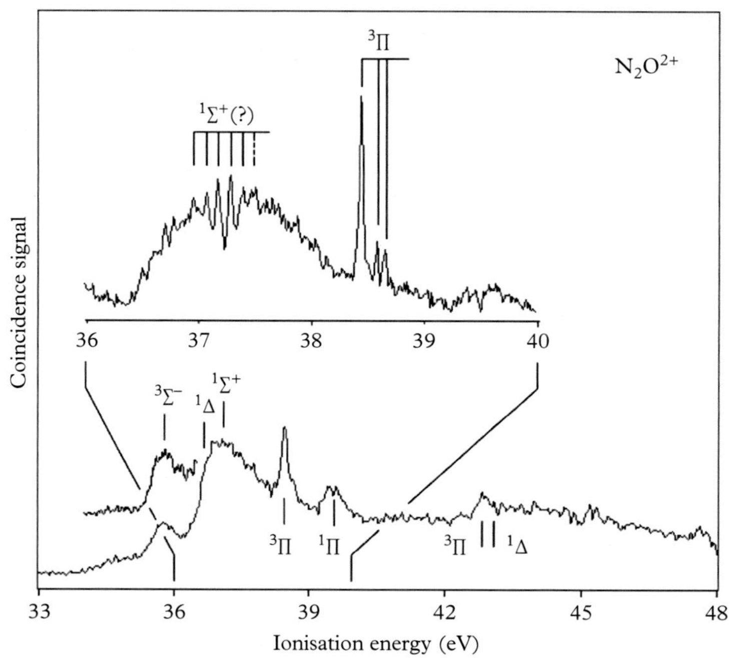

Figure 3.5.1 *TOF-PEPECO spectrum of N_2O at 48.37 eV and the part of the spectrum taken at 40.81 eV showing vibrational structure. No such structure is seen in the band at the lowest energy. The short section above the lower trace shows the effect of excluding low-energy electrons (KE < 2 eV) which come from dissociative autoionisation. The calculated energies of the lowest states and of two higher ones are marked [2].*

are consistent with this finding, and show a deep well with high barriers to dissociation for the $^3\Pi$ state. The ions in this state are found to emit fluorescence in falling down to the $X^3\Sigma^-$ state, which subsequently dissociates to $NO^+ + N^+$, some ions having a lifetime of a few hundred nanoseconds [4]. Evidence confirming the energies of the singlet states comes from the Auger spectra [5] and from double charge transfer [6], while the energies of the triplet states have been measured in separate DCT experiments [7]. The dynamics of dissociation in N_2O^{2+}, in particular the coupling of electron and nuclear motion on the shortest timescale, have been studied by Zhou et al. [8] using high harmonic generation of EUV light from a laser source.

The coincidence map of N_2O double photoionisation, Fig. 3.5.2, exhibits a new phenomenon in the form of a broad 'curtain' of dissociative double ionisation below the molecular double ionisation threshold.

In Fig. 3.5.2 there are several vertical ridges of intensity extending below and above the onset of molecular double ionisation (near 35.4 eV) and also some sloping lines within the higher ionisation energy range. These are all lines from atomic autoionisation, including recognisable electron energies from both O* and N* superexcited atoms, resulting from the same process as is seen so clearly in oxygen double ionisation (Chapter 2, Section 2.6). The new feature is the extended area of intensity below the molecular DIE forming a continuous 'curtain' and extending up to a sharp limit in electron energy of 1.9 eV which is the excitation energy of N^+ (^1D) above the N^+ ground state (^3S). It is

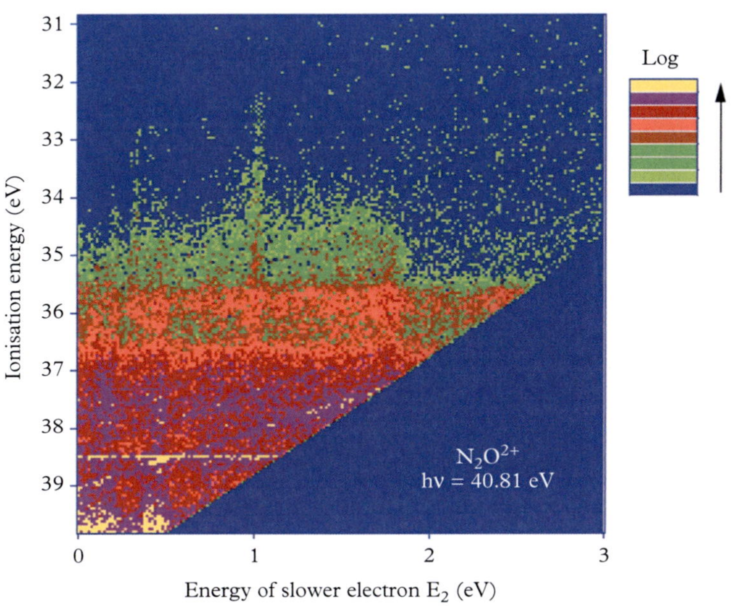

Figure 3.5.2 *Coincidence map for N_2O at 40.81 eV photon energy including all electrons and showing strong dissociative double ionisation below the molecular double ionisation energy of about 35.4 eV (see text).*

thought that the process involved is formation of superexcited N^* in one or many Rydberg states converging to this limit, which then autoionise at a range of distances from the NO^+ ion, so releasing electrons with a range of energies. The same phenomenon has been observed in the coincidence maps for NO_2 (Section 3.10) and a few other small molecules [9], but is not fully understood.

Double photoionisation of N_2O

State(s)	Ionisation energy (eV)
$X^3\Sigma^-$	Onset ~35.4, peak 35.9
$a^1\Delta$ or $b^1\Sigma^+$	Strong lines at 37.18 and 37.29
$A^3\Pi$	38.45
$c^1\Pi$	39.6

References

[1] A.W. Potts and T.A. Williams, *J. Electron Spectrosc. Related Phenom.*, **3** (1974) 3.
[2] S. Taylor, J.H.D. Eland and M. Hochlaf, *J. Chem. Phys.*, **124** (2006) 204319.
[3] N. Levasseur and P. Millié, *J. Chem. Phys.*, **92** (1990) 2974.
[4] T. Field and J.H.D. Eland, *Chem. Phys. Lett.*, **211** (1993) 436.
[5] W.J. Griffiths et al., *J. Phys. B: At. Mol. Opt. Phys.*, **24** (1991) 4187.
[6] S.D. Price et al., *J. Chem. Phys.*, **88** (1988) 1511.
[7] F.M. Harris et al., *J. Chem. Soc. Faraday Trans.*, **87** (1991) 1681.
[8] X. Zhou et al., *Nature Physics*, **8** (2012) 232.
[9] J.H.D. Eland, *Adv. Chem. Phys.*, **141** (2009) 103.

3.6 Carbonyl sulphide, OCS

Carbonyl sulphide is a linear molecule with the valence electron configuration

$$S2s^2S2p^66\sigma^27\sigma^28\sigma^29\sigma^22\pi^43\pi^4 \ldots\ldots\ldots\ldots\ldots\ldots\ldots {}^1\Sigma^+.$$

The binding energies of the four outermost orbitals are about 11.2, 15.5, 16 and 18 eV, according to the photoelectron spectrum [1], where only the 2π orbital appears as strongly bonding. Two calculations of the molecular electronic structure of OCS^{2+} [2, 3] are reasonably concordant in predicting the existence of about 15 electronic states in the binding energy range below 40 eV. The spectrum in Fig. 3.6.1 contains seven clearly distinguishable features in this range.

The three lowest energy bands in the spectrum (Fig. 3.6.1) are all ascribed to $3\pi^{-2}$ ionisation. Of these, only the second shows clearly resolved vibrational structure. The lack of observable structure in the $X^3\Sigma^-$ and $b^1\Sigma^+$ bands is rather surprising as they have the same electron configuration as the $a^1\Delta$ state and dissociation starts only at 33.5 eV [4]. The lack of detected structure must be due to spectral congestion, possibly caused by excitation of more than one vibrational mode or by spin splitting between the $\Omega = 0$ and $\Omega = \pm1$ sublevels in $X^3\Sigma^-$. In the $a^1\Delta$ state, there is a clear progression in the C—S stretching mode (ν_3); the vibration interval agrees well with the expected frequency [3]. Selective excitation of this particular vibration mode is probably related to

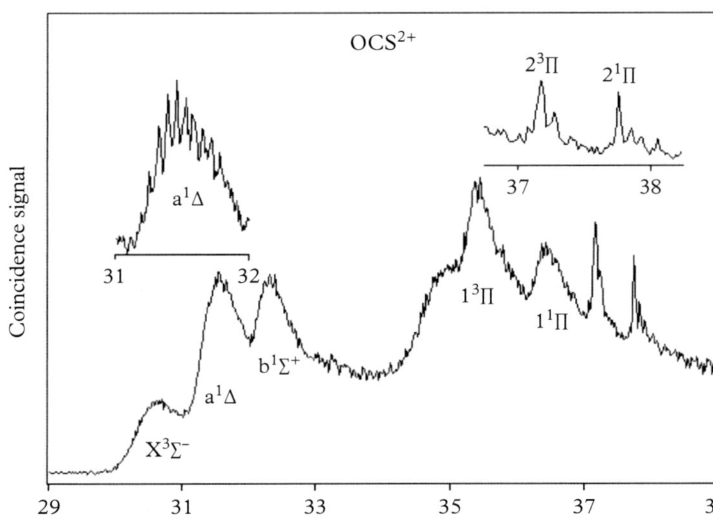

Figure 3.6.1 *TOF-PEPECO spectra of OCS taken with 40.81 eV photons (main trace) and upper right (37–38 eV insert) and at 32.69 eV (31–32 eV insert).*

the strong localisation of the 3π orbital on the S atom [5]. As in the spectra of the valence-isoelectronic molecules CO_2 and N_2O, states involving ionisation from the bonding 2π orbital produce only broad unstructured bands which contribute mainly to an underlying continuum. At higher energy, two pairs of bands can be rather confidently attributed on the basis of the calculations [3] to taking one electron from 3π and one from 9σ or 8σ. The difference in band shape between the two pairs ($1^3\Pi$ and $1^1\Pi$ unstructured, $2^3\Pi$ and $2^1\Pi$ resolved with prominent (0—0) transitions) is probably related to avoided crossings between the potential surfaces along the C—S stretching coordinate [3]. Energies of the two lowest states of OCS^{2+} were measured by TPEsCO [6] and agree well with earlier measurements by double charge exchange [7, 8]. Some higher energy states, presumably singlets, have been located in the Auger spectra [5] and ascribed to particular dominant electron configurations. There has also been recent work on the dissociations of OCS^{2+} [9] and on the triple ionisation of the molecule [10].

States of OCS^{2+} (with adiabatic DIEs as bold)

State(s)	DIE (eV)	Orbitals ionised
$X^3\Sigma^-$	**30.0** (30.6)	$3\pi^{-2}$
$a^1\Delta$	**31.1** (31.46)	$3\pi^{-2}$
$b^1\Sigma^+$	32 (32.3)	$3\pi^{-2}$
$1^3\Pi$	35.4	$9\sigma^{-1}3\pi^{-1}$
$1^1\Pi$	36.5	$9\sigma^{-1}3\pi^{-1}$
$2^3\Pi$	**37.18**	$8\sigma^{-1}3\pi^{-1}$
$2^1\Pi$	**37.76**	$8\sigma^{-1}3\pi^{-1}$

References

[1] D.M.P. Holland and M.A. MacDonald, *Chem. Phys.*, **144** (1990) 279.
[2] P. Millié et al., *J. Chem. Phys.*, **84** (1986) 1259.
[3] V. Brites et al., *Chem. Phys.*, **346** (2008) 23.
[4] P. Lablanquie et al., *Photophysics and Photochemistry above 6 eV*, ed. F. Lahmani, Elsevier, Amsterdam, 1985, p. 25.
[5] T.X. Carroll, D. Ji and T.D. Thomas, *J. Electron Spectrosc. Related Phenom.*, **51** (1990) 471.
[6] R.I. Hall et al., *Chem. Phys.*, **187** (1994) 125.
[7] W.G. Griffiths and F.M. Harris, *Int. J. Mass Spectrom. Ion Processes*, 87 (1989) 349.
[8] M.L. Langford et al., *Chem. Phys.*, **149** (1991) 445.
[9] P. Wang and C.R. Vidal, *J. Chem. Phys.*, **118** (2003) 5383.
[10] J.H.D. Eland et al., *J. Chem. Phys.*, **132** (2010) 014311.

3.7 Carbon disulphide, CS_2

Carbon disulphide is linear and symmetric as SCS and is isoelectronic with carbon dioxide in the valence shell, so the relevant part of its electron configuration outside the S inner shells can be written as

$$5\sigma_g{}^2 4\sigma_u{}^2 6\sigma_g{}^2 5\sigma_u{}^2 2\pi_u{}^4 2\pi_g{}^4 \ldots\ldots\ldots\ldots\ldots\ldots\ldots {}^1\Sigma_g{}^+.$$

Here we have left the 1s shell of carbon and the 1s, 2s and 2p shells of sulphur out of the count to maintain consistency with recent publications. The binding energies of the outermost orbitals are near 10.1, 12.7, 14.5 and 16.2 eV, as seen in the photoelectron spectrum [1]. The $2\pi_g$ orbital is non-bonding and located purely on the sulphur atoms; only the $2\pi_u$ orbital shows any appreciable bonding character. High-level calculations [2, 4] predict the existence of about 20 states of the $CS_2{}^{2+}$ ion below 37 eV, and the Auger spectrum [5] from vacancy creation in S2p suggests that there are about five

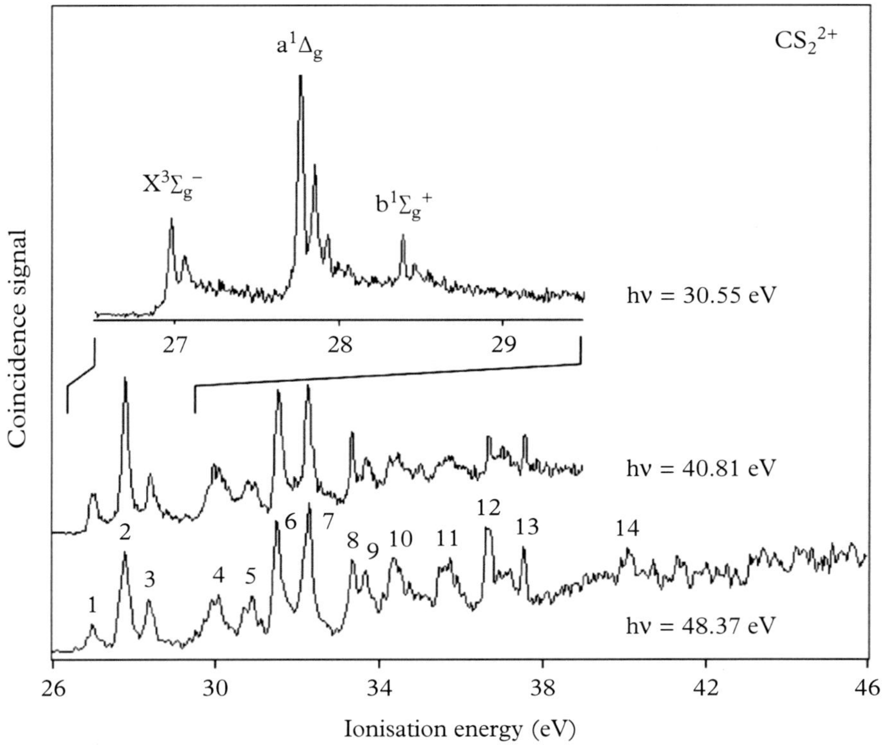

Figure 3.7.1 *TOF-PEPECO spectra of CS_2 taken with ionising light of the HeIIα (40.81 eV) and HeIIβ (48.37 eV) lines and with the higher energy component of the NeII doublet near 406 Å (30.55 eV).*

more singlet states below 40 eV. The TOF-PEPECO spectrum has been recorded at several photon energies, as shown in Fig. 3.7.1.

The first three states in the spectrum arise in the familiar way from removal of two electrons from the non-bonding $2\pi_g$ orbital and were first resolved by the TPEsCO technique [6]. The vibrational structure seen in that work was more complex than the simple progressions in the symmetric stretching vibration (v_1) evident in the TOF-PEPECO spectrum [7], probably because of some interfering autoionisation effects. The lowest photon energy at which dissociation is observed is near 30 eV [4, 8], so levels in the state(s) represented by the fourth band in Fig. 3.7.1 may also be stable. It turns out that fluorescence is observed from the CS_2^{2+} ions at binding energies of 29.7 eV (in band 4) and at 30.7 eV (in band 5). The emitting state in band 4 has been identified as $1^3\Sigma_u^-$, which is one of a number of states which may make up this band. The emitter in band 5 is the $^1\Delta_u$ state, which like $1^3\Sigma_u^-$ arises from $2\pi_u^{-1}2\pi_g^{-1}$ ionisation and is the only state which the calculations locate in this band [4]. Neither band 4 nor band 5 shows any resolved vibrational structure, even when the ionisation is by 32.69 eV photons, allowing high resolution. It seems probable that spectral congestion is the cause of this lack of resolvable structure, as it is likely that at least some states with the $2\pi_u^3 2\pi_g^3$ electron configuration have non-linear equilibrium geometries, causing excitation of the low-frequency bending vibration.

Identifications of the higher energy states of CS_2^{2+} suggested below are guided mainly by the numerical calculations [2–4], but also by the Auger spectra [5] and DCT results [9], and by comparison with the double ionisation spectra of OCS.

States of CS_2^{2+} (with adiabatic DIEs as bold)

Band	State(s)	DIE (eV)	Orbitals ionised
1	$X^3\Sigma_g^-$	**26.98**	$2\pi_g^{-2}$
2	$a^1\Delta_g$	**27.76**	$2\pi_g^{-2}$
3	$b^1\Sigma_g^+$	**28.39**	$2\pi_g^{-2}$
4	$1^3\Sigma_u^-$ + others	30.0	$2\pi_u^{-1}2\pi_g^{-1}$
5	$1^1\Delta_u$	30.8	$2\pi_u^{-1}2\pi_g^{-1}$
6	$^3\Pi_u$	31.52	$5\sigma_u^{-1}2\pi_g^{-1}$
7	$^1\Pi_u$	32.28	$5\sigma_u^{-1}2\pi_g^{-1}$
8	$^3\Pi_g$	33.33	$6\sigma_g^{-1}2\pi_g^{-1}$
9	$^1\Pi_g$	33.65	$6\sigma_g^{-1}2\pi_g^{-1}$
10	$^3\Pi_u$	34.4	$6\sigma_g^{-1}2\pi_u^{-1}$
11	$^1\Pi_u$	35.6	$6\sigma_g^{-1}2\pi_u^{-1}$
12	^{1}Let	36.7	
13	??	37.6	
14	^{1}Let	40.1	

References

[1] P. Baltzer et al., *Chem. Phys.*, **202** (1995) 2168.
[2] P. Millié et al., *J. Chem. Phys.*, **84** (1986) 1259.
[3] M. Hochlaf, G. Chambaud and P. Rosmus, *J. Chem. Phys.*, **108** (1998) 4047.
[4] S. Taylor, J.H.D. Eland and M. Hochlaf, *Chem. Phys.*, **330** (2006) 16.
[5] U. Alkemper and F. von Busch, *J. Electron Spectrosc. Related Phenom.*, **93** (1998) 115.
[6] M. Hochlaf et al., *Chem. Phys.*, **234** (1998) 249.
[7] J.H.D. Eland, *Chem. Phys.*, **294** (2003) 171.
[8] P. Lablanquie et al., *J. Chem. Phys.*, **82** (1985) 2951.
[9] M.L. Langford et al., *Chem. Phys.*, **149** (1991) 445.

3.8 Bromine and iodine cyanides, BrCN and ICN

The cyanogen halides (halogen cyanides) are linear molecules, isoelectronic in the valence shells with carbon dioxide and the related triatomic compounds. The order of the orbitals as seen in the photoelectron spectra [1, 2] is the same in BrCN and ICN, and we write it without account of the deeper inner shells as

$$1\sigma^2 2\sigma^2 1\pi^4 3\sigma^2 2\pi^4 \ldots\ldots\ldots\ldots\ldots\ldots\ldots\ldots\ldots\ldots {}^1\Sigma^+.$$

In BrCN the binding energies of the outer orbitals are roughly 12, 13.6, 14.5 and 18 eV while in ICN they are 11.3, 13.2, 13.6 and 17 eV. In double ionisation spectra, we can expect to see the usual set of states, possibly modified by the localisation of the very strong bonding between C and N and weak bonding to the halogen and by spin-orbit coupling in otherwise degenerate states. As there are no Auger spectra, TPEsCO spectra or wide-range high-level molecular structure calculations known to us for these molecules, we rely mainly on the orbital energies and comparison with other spectra to interpret the TOF-PEPECO results [3]. Detailed calculations have been done for some of the individual resolved states, emphasising the role of spin-orbit effects [4].

In the spectrum of $BrCN^{2+}$ (Fig. 3.8.1) the resolved structure as seen at the lowest ionisation energies disappears above 33 eV, which is the measured onset of dissociation

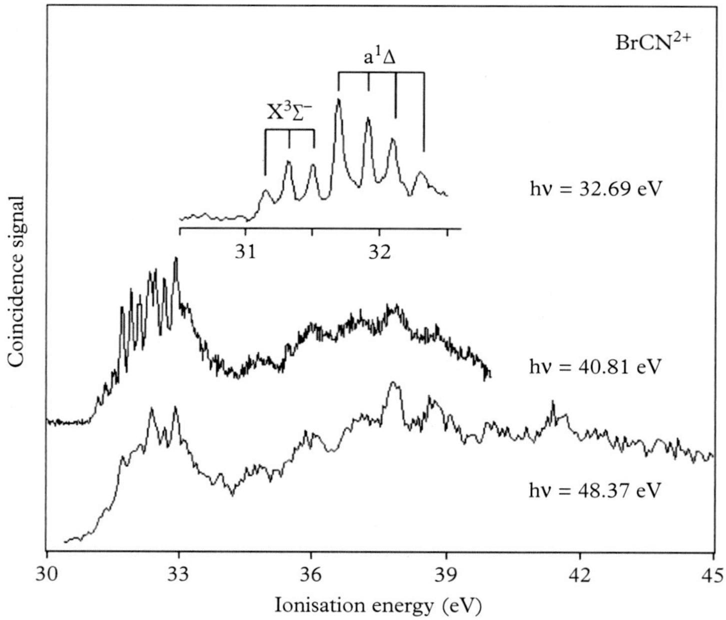

Figure 3.8.1 *TOF-PEPECO spectra of BrCN at the three photon energies indicated.*

to Br^+ + CN^+ [3]. The first two states from $2\pi^{-2}$ ionisation are clearly resolved in the spectrum taken at 32.69 eV, and show excitation of the C—N stretching vibration. The following lines between 32 and 33 eV must be partly attributed to the expected $^1\Sigma^+$ state from $2\pi^{-2}$, but must also include the $^3\Pi$ state from $3\sigma^{-1}2\pi^{-1}$ to account for the irregular line intensities. The relativistic calculations [4] indicate that $^3\Pi$ is significantly split by spin–orbit interaction, making the interpretation of the lines in this region very complex. The $^1\Pi$ state from $3\sigma^{-1}2\pi^{-1}$ probably lies close by, where the intensity falls off between 33 and 34 eV. The rounded bands between 34 and 40 eV cannot be assigned without proper calculations, but probably come from $2\sigma^{-1}2\pi^{-1}$, $3\sigma^{-2}$ and $1\pi^{-1}2\pi^{-1}$ ionisations. On the basis of the orbital energies, we can tentatively attribute the rather isolated band at 41.5 eV to the $^1\Sigma^+$ state from $2\sigma^{-2}$. The dissociations of $BrCN^{2+}$ were studied in some detail as part of the TOF-PEPECO work [3] and have also been studied at some higher photon energies [5].

As for BrCN, the double photoionisation spectrum of ICN (Fig. 3.8.2) shows a loss of resolution where dissociation of ICN^{2+} to I^+ + CN^+ sets in above 31 eV [3]. The broadening of all spectral bands at higher ionisation energy is evident in the spectrum. As for BrCN, the two lowest energy states of the dication are well resolved at the lower photon energy, and all other states occur as overlapping bands. In the $X^3\Sigma^-$ state, what appears to be a simple progression in the C—N stretching vibration is actually more complex, because the calculated spin-orbit splitting of 125 meV between the Ω=0 and $\Omega = \pm1$ components of this state is close to the measured peak spacing of 200±10 meV.

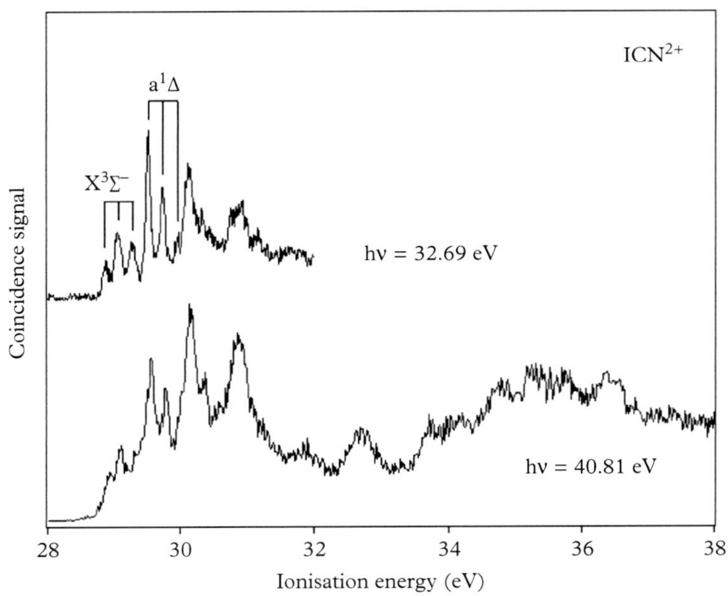

Figure 3.8.2 *TOF-PEPECO spectra of ICN at two photon energies.*

The second peak of the apparent progression is undoubtedly composite. The same spacing, with a more normal Frank–Condon envelope, is seen for the non-degenerate $^1\Delta$ state. After $^1\Delta$, two rather intense broad bands are seen at 30.14 and 30.85 eV; although these are close to the onset of dissociation, their broadness is probably to be attributed more to overlapping of multiple lines. We believe that major components of these bands are the $^3\Pi$ and $^1\Pi$ states from $3\sigma^{-1}2\pi^{-1}$. The broad bands at higher energies cannot be assigned to particular states at present, but the general resemblance to the double ionisation spectrum of BrCN is apparent.

Identified and unidentified states of $BrCN^{2+}$ (with adiabatic DIEs as bold)

State(s)	DIE (eV)	Orbitals ionised
$X^3\Sigma^-$	**31.11**	$2\pi^{-2}$
$a^1\Delta$	**31.66**	$2\pi^{-2}$
?	32.26	
?	32.39	
?	32.61	
?	32.85	

Identified and unidentified states of ICN^{2+} (with adiabatic DIEs as bold)

State(s)	DIE (eV)	Orbitals ionised
$X^3\Sigma^-$	**28.88**	$2\pi^{-2}$
$a^1\Delta$	**29.52**	$2\pi^{-2}$
?	30.13	
?	30.34	
?	30.87	
?	32.6	

References

[1] J.H.D. Eland et al., *Chem. Phys.*, **212** (1996) 457.
[2] J.H.D. Eland et al., *Chem. Phys.*, **222** (1997) 229.
[3] J.H.D. Eland and R. Feifel, *Chem. Phys.*, **327** (2006) 85.
[4] S. Manna and S. Mischra, *J. Phys. Chem. A*, **120** (2016) 1554.
[5] T. Ibuki et al., *Chem. Phys.*, **203** (1996) 359.

3.9 Mercuric chloride, $HgCl_2$

Mercuric chloride is a linear symmetric molecule, ClHgCl, which, as its old name of 'corrosive sublimate' implies, is relatively volatile, sufficiently so as to be measured as vapour in simple spectrometers. Its valence orbital configuration is augmented by the presence of the 5d shell in the same energy range, interacting with the valence electrons. We write its configuration outside the inner shells as

$$5d_{3/2}{}^4 5d_{5/2}{}^6 1\sigma^2 2\sigma^2 1\pi^4 2\pi^4 \ldots\ldots\ldots\ldots\ldots\ldots\ldots {}^1\Sigma^+.$$

The 5d shell is split widely by spin–orbit interaction and less widely by interaction with the molecular field. The energies of the σ- and π-orbitals from the photoelectron spectrum [1–3] are 11.4, 12.1, 12.7 and 13.7 eV and the 5d ionisations are near 17 and 18.5 eV. The change in relative intensities of the photoelectron bands between 21.22 and 40.81 eV photon energy is notable. The spectrum in Fig. 3.9.1 is, as far as we know,

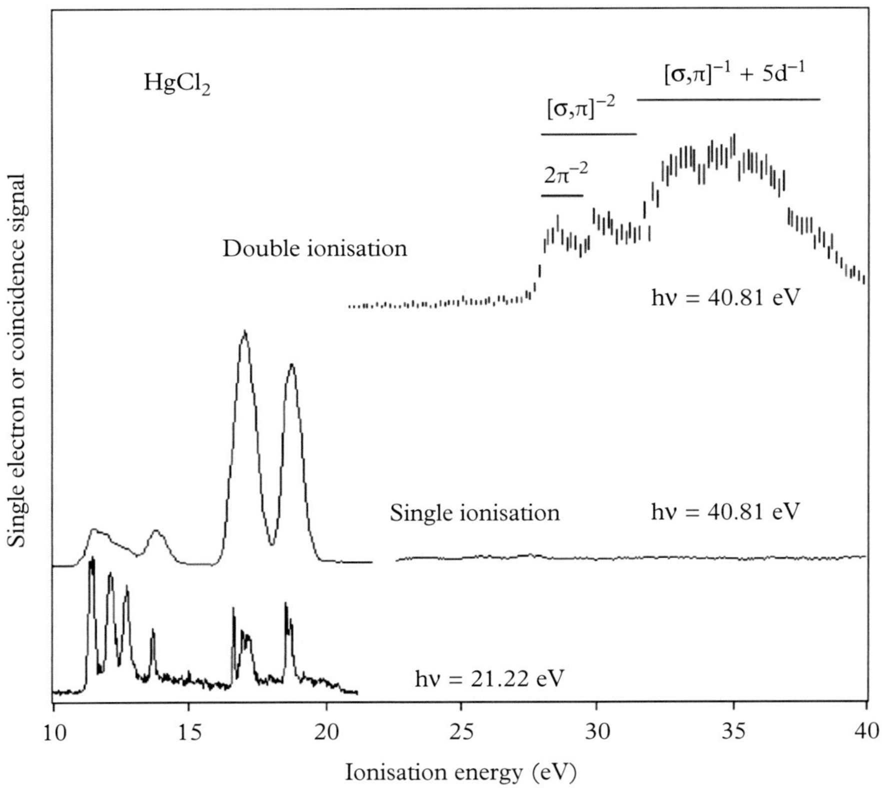

Figure 3.9.1 *TOF-PEPECO spectrum of $HgCl_2$ at 40.81 eV photon energy and single ionisation photoelectron spectra taken at two photon energies using the same apparatus.*

the only double ionisation spectrum of this molecule that has been measured. To show the importance of the 5d shell, we include both single and double ionisation in Fig. 3.9.1.

The double ionisation spectrum is not resolved into individual states, almost certainly because there are many overlapping electronic states within a narrow energy range, as predicted by combinations of the orbital energies. The best we can do is to indicate the major configurations of the states in each region, as indicated in the figure. The onset of the first band is at 27.9±0.2 eV and its peak is near 28.5 eV.

References

[1] J.H.D. Eland, *J. Mass Spectrom. Ion Phys.*, **4** (1970) 37.
[2] J. Berkowitz, J. *Chem. Phys.*, **61** (1974) 407.
[3] R. Egdell et al., *J. Electron Spectrosc. Related Phenom.*, **12** (1977) 415.

3.10 Nitrogen dioxide, NO_2

Nitrogen dioxide is a non-linear but symmetric molecule, ONO, with a bond angle in the ground state of 134°. When it is ionised either singly or doubly the angle changes, causing excitation of long progressions in the bending vibration. Each orbital of degenerate π-character in the related linear molecules here becomes two non-degenerate orbitals. NO_2 is also an open-shell molecule. We write its configuration, including the seven orbitals with ionisation energies below 30 eV, as

$$4a_1{}^2 3b_2{}^2 1b_1{}^2 5a_1{}^2 1a_2{}^2 4b_2{}^2 6a_1{}^1 \ldots\ldots\ldots\ldots\ldots\ldots\ldots {}^2A_1.$$

Because of the unpaired spin, the bending vibrations and the large number of orbitals, even its single ionisation spectrum [1] is highly complex and we cannot expect to resolve much structure in the double ionisation spectrum; indeed, we see no resolved states. What is found is a quite unusually strong contribution from dissociative double ionisation involving superexcited oxygen atoms, as shown by the coincidence map in Fig. 3.10.1.

The numerous vertical lines on the map all represent superexcited states of atomic oxygen, formed by dissociation of intermediate NO_2^+ monocations, which later autoionise, finally populating the ground state of O^+, $^4S_{3/2}$. They extend to a very low double ionisation energy, not far above the thermodynamic limit for formation of O^+ + NO^+ in their ground states at 26 eV. Although it is hard to see in this map, the autoionisation lines persist above the DIE right up to the photon energy. They are less prevalent when the molecule is ionised at 48.37 eV instead. The 'curtain' of intensity at ionisation energies below the molecular double ionisation threshold probably has essentially the same origin here as in the case of N_2O (Section 3.5); it extends up to 3.3 eV, which

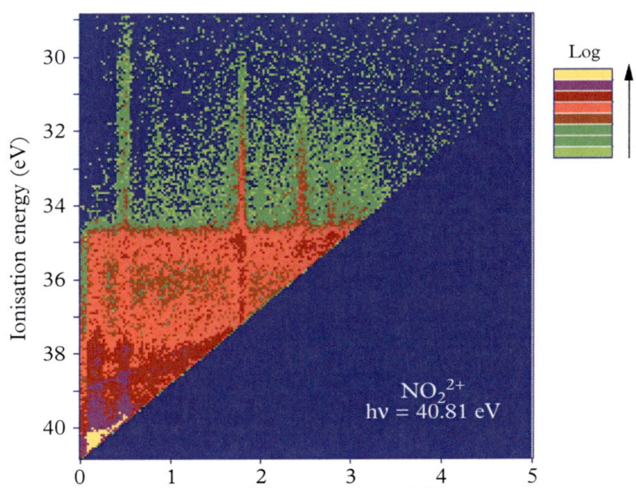

Figure 3.10.1 *Coincidence map for double ionisation of NO_2 at 40.81 eV.*

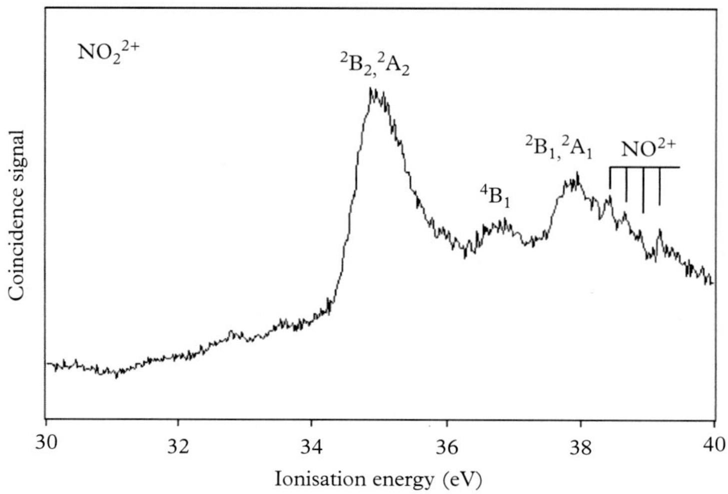

Figure 3.10.2 *TOF-PEPECO spectrum of NO_2 at 40.81 eV. Impurity NO^{2+} bands are marked.*

is the energy of the first excited state (term) of O^+, the ^2D state. Our supposition is that it represents multiple Rydberg states converging to the $O^+(^2D)$ state, which decay by autoionisation at different distances from the NO^+ partner ion. There are so many dissociative autoionisation pathways that it is not possible in this case to eliminate them and offer a spectrum of pure molecular double photoionisation of NO_2. The spectrum in Fig. 3.10.2 thus includes all electrons and shows a strong rising background.

Another problem with NO_2 is that it is very difficult to remove all traces of NO impurity, and this was not achieved for the spectrum above. The spectrum of $NO_2{}^{2+}$ has previously been investigated by a combination of Auger spectroscopy, photodissociation, double charge exchange and high-level theory [2]. The lowest energy state of the dication at the geometry of neutral NO_2 is reached by removing the outermost (lone) electron from $6a_1$ and one electron from $4b_2$, but at equilibrium in the resulting state the orbitals $1a_2$ and $4b_2$ become degenerate as π_g and the molecular dication is linear. According to the calculations, linear geometry is not reached within the Franck-Condon zone, so the apparent onset of the first band at 34.3±0.1 eV cannot be reliably interpreted as an adiabatic ionisation energy. The peak of the first composite band is at 35 eV, the second at 36.8 eV and the third at 37.9 eV; the state identifications shown are derived from the calculations [2], with which the band maxima agree well. The fragmentations of $NO_2{}^{2+}$ have been examined in some detail in related work [3, 4].

References

[1] P. Baltzer et al., *Chem. Phys.*, **237** (1998) 451.
[2] P.G. Fournier et al., J. *Chem. Phys.*, **89** (1988) 3553.
[3] J.H.D. Eland, *Mol. Phys.*, **61** (1987) 725.
[4] T. Masuoka and A. Kobayashi, *Chem. Phys.*, **302** (2004) 31.

3.11 Sulphur dioxide, SO_2

Like nitrogen dioxide, sulphur dioxide, OSO, is non-linear but symmetric with a bond angle in the ground state of 118°. Its outer molecular orbital line-up is like that of NO_2 in terms of symmetry, but we write its configuration with numbering which includes the extra S2p orbitals, for consistency with other sources:

$$6a_1{}^2 3b_2{}^2 2b_1{}^2 7a_1{}^2 4b_2{}^2 5b_2{}^2 1a_2{}^2 8a_1{}^2 \ldots\ldots\ldots\ldots\ldots\ldots\ldots {}^1A_1.$$

Its photoelectron spectrum has been examined at high resolution and over a wide energy range, and contains complex and overlapping structures [1]. The three outermost bands are centred at about 12.5, 13.2 and 13.5 eV. In double photoionisation there is some dissociative autoionisation via atomic oxygen seen in the coincidence map, but it is quite minor in this case and does not significantly interfere with the double ionisation spectrum.

As expected, many states contribute to the valence double ionisation of this molecule and most bands are highly congested because of excitation of the low-frequency bending vibration on ionisation. The spectrum had been explored by Auger spectroscopy [2] and by double charge transfer [3] before the first detailed TOF-PEPECO investigation [4], and these studies were accompanied by extensive theoretical work. The highest-level

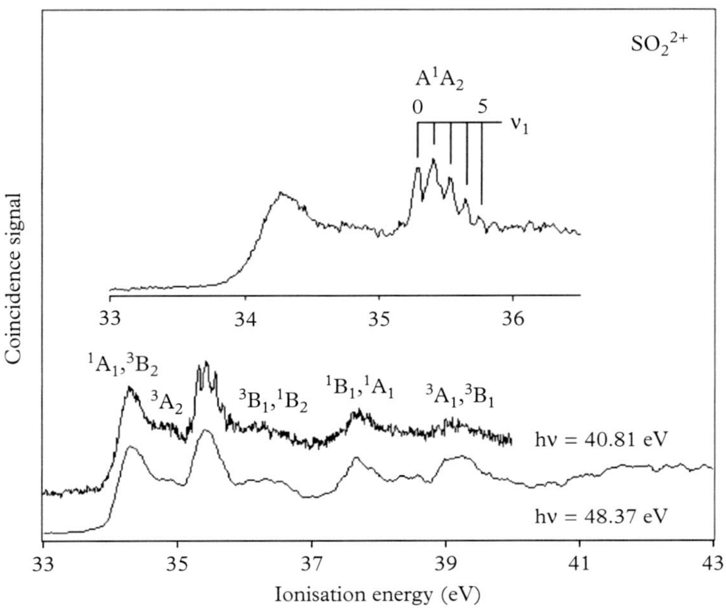

Figure 3.11.1 *TOF-PEPECO spectra of SO_2 at 48.37 and 40.81 eV photon energies, with a blow-up of the low-energy section.*

theory was done to interpret the latest TOF-PEPECO work [4] and is the basis for the current interpretation of the spectrum shown as Fig. 3.11.1.

If two electrons are removed from the outermost orbital, the doubly charged ion thus produced is valence-isoelectronic with CO_2 and would be linear at equilibrium. The linear configuration is so far from the Franck–Condon zone in SO_2 that it is not even approached, and the $^1A_1$ state of SO_2^{2+} simply contributes to the first band in the spectrum, whose peak is at 34.3 eV. Theory predicts that the linear ground state lies near 33.5 eV [4]. There is one vibrationally resolved band with adiabatic ionisation energy of 35.28 eV and a principal vibrational spacing of 130 meV corresponding to the symmetric S—O stretching vibration. Comparison with the DCT results and the latest calculations allows this to be identified as the $^1A_2$ state from $1a_2^{-1}8a_1^{-1}$ ionisation. The other identifications of bands with states as given in the figure are somewhat less certain, and are based mainly on the calculations.

Besides the spectrum, the fragmentation pathways of doubly ionised SO_2 have been extensively investigated by related coincidence methods [5–7].

References

[1] D.M.P. Holland et al., *Chem. Phys.*, **188** (1994) 317.
[2] M. Thompson, P.A. Hewitt and D.A. Wooliscroft, *Anal. Chem.*, **48** (1976) 1336.
[3] I.W. Griffiths, D.E. Parry and F.M. Harris, *Chem. Phys.*, **238** (1998) 21.
[4] M. Hochlaf and J.H.D. Eland, *J. Chem. Phys.*, **120** (2004) 6449.
[5] G. Dujardin et al., *Chem. Phys.*, 88 (1984) 339.
[6] S. Hsieh and J.H.D. Eland, *J. Phys. B: At. Mol. Opt. Phys.*, **30** (1997) 4515.
[7] T.A. Field and J.H.D. Eland, *Int. J. Mass Spectrom. Ion Processes*, **192** (1999) 281.

Chapter 4

Molecules with four, five or seven atoms

4

Molecules with four, five or seven atoms

There are relatively few important molecules with just four atoms, but most of the ones included here are present and sometimes abundant in planetary and astrophysical environments. The range of five-atom molecules includes methane and all its simple derivatives, of which we present a selection. Where possible we group closely related molecules together in this chapter, as much of the discussion of their electronic structure is the same for all members of a group. Ordering of the molecules by the number of atoms followed by valence electron count and then by molecular weight is loosely followed within each group. We also include SF_6 as a closely related molecule even though its atom count goes beyond those of some molecules in later chapters.

The structures given for the molecules in this and other chapters, used as the basis of the orbital symmetries, are the conventional *time-averaged* equilibrium structures. Where there are degenerate vibration modes, as in linear and high-symmetry molecules, the most probable *instantaneous* structure may be different and of lower symmetry. Some molecules are 'floppy', having low-frequency bending vibrations which can be excited thermally at room temperature, or 'fluxional', meaning that thermal energy or even zero-point energy may allow individual atoms to exchange positions. Because ionisation and double ionisation by photon impact can be considered instantaneous on the timescale of molecular vibration, the reduced symmetry may have consequences for the spectra and certainly for the dissociation processes which follow ionisation. Detailed consideration of such effects is a complication too far for almost all theoretical treatments of doubly charged ions and for this book, but if future work achieves higher specificity, accuracy and resolution, they will have to be borne in mind.

Double Photoionisation Spectra of Molecules. John H. D. Eland and Raimund Feifel.
© John H. D. Eland and Raimund Feifel, 2018. Published in 2018 by Oxford University Press.
DOI 10.1093/oso/9780198788980.001.0001

4.1 Ammonia, NH_3

The electronic structure of the pyramidal ammonia molecule in C_{3v} symmetry is

$$N1s^2N2s^21e^43a_1{}^2 \ldots\ldots\ldots\ldots\ldots\ldots {}^1A_1.$$

From the photoelectron spectrum [1], the binding energies of electrons in the two outer orbitals are about 10.5 and 16 eV, so if the Coulomb energy is about 10 eV the double ionisation spectrum is expected to contain a first band at about 30 eV (2 × 10.5 + 10), followed by two bands around 36 eV (10.5 + 16 + 10), with several more at higher energies. The removal of a single electron from the $3a_1$ orbital causes the molecule to become planar at equilibrium, with consequent excitation of a long progression in the 'umbrella' bending vibration seen in the photoelectron spectrum, and a similar change of geometry must be expected when two electrons are removed.

The lowest energy state of $NH_3{}^{2+}$ comes from removal of two electrons from the outermost $3a_1$ orbital, giving a ${}^1A_1$ state, which would be planar at equilibrium. The onset of the first band seen in Fig. 4.1.1, at 33.4 eV, may or may not represent formation of the molecular dicationic ground state, but is probably close to the adiabatic double ionisation energy. When a single electron is removed by photoionisation from the $3a_1$ orbital, the photoelectron spectrum shows the adiabatic transition, though weakly, and the structural change in $3a_1{}^{-2}$ ionisation must be similar. On the other hand, the absence of any resolvable fine structure in the first band must almost certainly be ascribed to

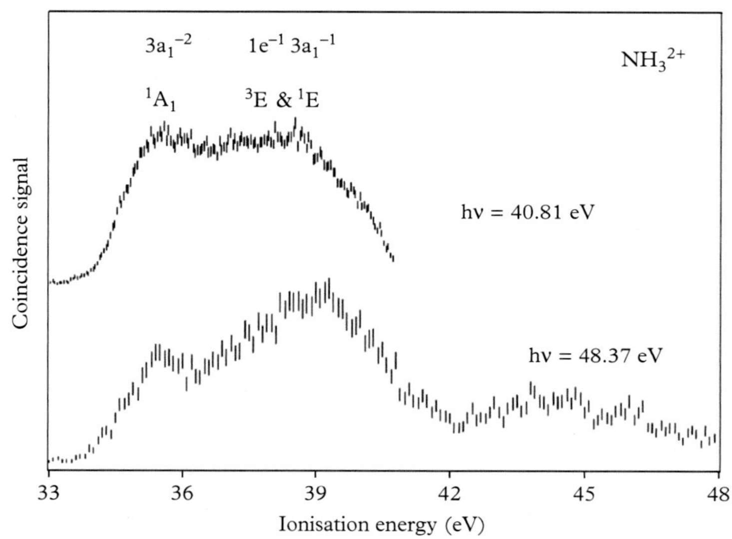

Figure 4.1.1 *TOF-PEPECO spectra of ammonia at the two HeII photon energies indicated. The signal intensities are here represented by ±1σ error bars.*

vibrational congestion because the onset of dissociation is at 35.5 eV [2], well above the start of the band, and stable NH_3^{2+} is found in the mass spectrum. To interpret the other bands in the TOF-PEPECO spectrum [3], we turn to previous investigations of doubly ionised ammonia by Auger spectroscopy [4, 5], double charge transfer (DCT) [6, 7] and numerical calculations [8]. The second main band, centred near 39 eV, certainly comes from $1e^{-}3a_1^{-1}$ ionisation and includes a ^3E state and a ^1E state, calculated to lie near 37.5 and 39 eV respectively. The high energy tail of the band may have a contribution from the $^3A_2$ state from $1e^{-2}$ ionisation, calculated at 43 eV. The other two states from the resulting $1e^2 3a_1^2$ configuration, ^1E and $^1A_1$, are probably both present in the band around 45 eV in Fig. 4.1.1. A single band at this same energy is seen in the Auger spectra [4, 5], which also show higher singlet states from configurations where an N2s electron is removed.

References

[1] J.W. Rabalais et al., *J. Chem. Phys.*, **58** (1973) 3370.
[2] M. Stankiewics et al., *J. Phys. B: At. Mol. Opt. Phys.*, **22** (1989) 21.
[3] J.H.D. Eland, *Chem. Phys.*, **323** (2006) 391.
[4] J.M. White, R.R. Rye and J.E. Houston, *Chem. Phys. Lett.*, **47** (1977) 146.
[5] R.W. Shaw, J.S. Jen and T.D. Thomas, *J. Electron Spectrosc. Related Phenom.*, **11** (1977) 91.
[6] P.G. Fournier et al., *J. Chem. Phys.*, **83** (1985) 241.
[7] W.J. Griffiths and F.M. Harris, *Rapid Commun. Mass Spectrom.*, **4** (1990) 366.
[8] F. Tarantelli et al., *Chem. Phys. Lett.*, **117** (1985) 577.

4.2 Acetylene (ethyne), C_2H_2

Acetylene is a symmetrical linear molecule, HCCH, with nominally the same valence electronic structure as the molecules N_2, CO and HCN. Its electron configuration is

$$C1s^4 2\sigma_g^{\ 2} 2\sigma_u^{\ 2} 3\sigma_g^{\ 2} 1\pi_u^{\ 4} .... {}^1\Sigma_g^{\ +}.$$

Removal of a single electron from the $1\pi_u$ orbital produces a photoelectron band near 11.4 eV, showing only a small change in the C—C bond length [1]. Ionisation from the $3\sigma_g$ orbital at about 17 eV causes more structural change and the photoelectron band is further congested by interaction with other states [2]. Stable ions with the formula $C_2H_2^{\ 2+}$ are observed in the mass spectrum of acetylene itself and of many other species. Some of them probably have the vinylidene structure CH_2C^{2+}, to which the acetylene dication can rearrange on a very short (<100 fs) timescale from its original $HCCH^{2+}$ form [3, 4].

The double ionisation of acetylene had been investigated in some detail by Auger spectroscopy [5], double charge transfer (DCT) [6–8] and TPEsCO [9] before the most detailed combined study which included TPEsCO, TOF-PEPECO and theory [10]. The DCT work of Furuhashi et al. [8] was the first to achieve vibrational resolution, revealing a progression in the C—C stretching vibration in the first singlet state, $a^1\Delta_g$. As with all the related molecules, removal of two electrons from the outermost π-orbital produces three states, $X^3\Sigma_g^{\ -}$, $a^1\Delta_g$ and $b^1\Sigma_g^{\ +}$. These three states are heavily

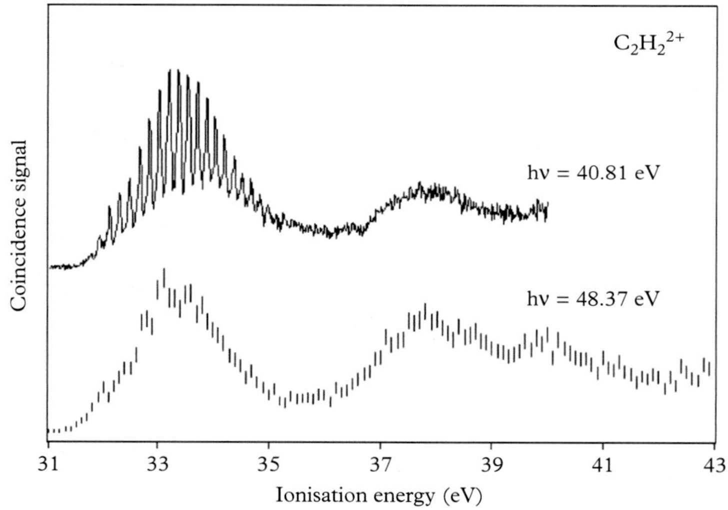

Figure 4.2.1 *TOF-PEPECO spectra of acetylene at two photon energies. The spectrum at 48.37 eV is represented by error bars because of the small number of counts available.*

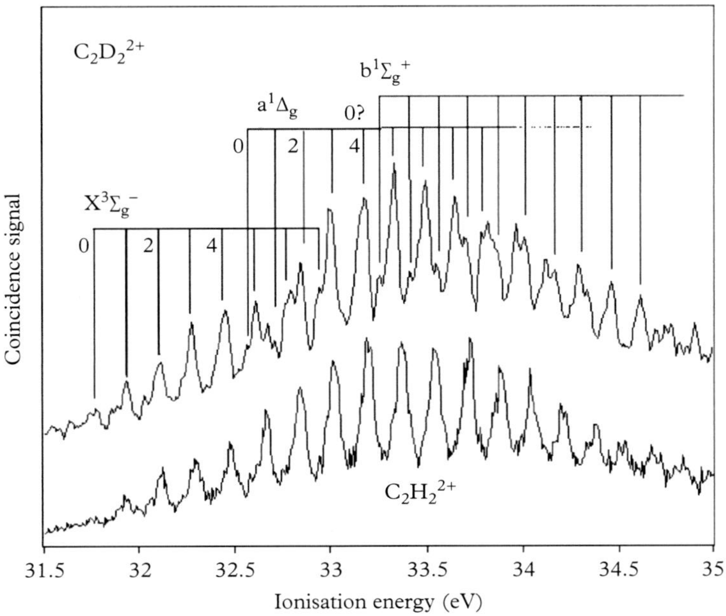

Figure 4.2.2 *Expanded views of the high resolution TOF-PEPECO spectra of acetylene and its fully deuterated analogue, showing the proposed decomposition of the band into contributions from the three lowest states. The vibration level numbering (0, 2, 4, etc.) for the symmetric C—C stretching mode is provisional.*

overlapping in the first band of the TOF-PEPECO spectrum, Fig. 4.2.1, and their entanglement could not be unravelled until measurements were made on the fully deuterated isotopomer C_2D_2.

When one electron is removed from $1\pi_u$ and one from $3\sigma_g$, we expect the two states $A^3\Pi_u$ and $d^1\Pi_u$; these two states are dissociative in the HCC—H bond-breaking coordinate [11] and all the calculations place them near 37 and 38 eV respectively. There is a broad band in the TOF-PEPECO spectrum in Fig. 4.2.1 taken at 40.81 eV with its peak at 38 eV, and both this and a second band near 40 eV appear in the spectrum taken at 48.37 eV. The 38 eV band certainly includes the $A^3\Pi_u$ state, but according to the DCT results [7] it is overlapped at 38.4 eV by $d^1\Pi_u$. The higher band may include other states calculated to lie in this range by Gaire et al. [11].

High resolution spectra of two isotopic forms of acetylene are shown in Fig. 4.2.2. The vibration excited by removing two of the π-electrons from C_2H_2 or C_2D_2 is the symmetric C—C stretch with almost constant vibration intervals of 173, 168 and 150 meV in the three states in the better-resolved C_2D_2 spectrum. Because of overlaps it is impossible to derive anharmonicities; the uncertainty in the intervals is estimated as ± 5 meV. The change in this vibration frequency between C_2H_2 and C_2D_2 is quite small (factor

0.96 from the change in reduced mass) but is just sufficient to make the peak positions different in the two isotopomers for the higher vibration levels in each state. This observation and the expectation of similar Franck–Condon envelopes are the basis of the proposed decomposition.

States of $C_2H_2^{2+}$ (energies in eV)

State	Adiabatic	Vertical	Ionisation
$X^3\Sigma_g^-$	31.76	32.5	$1\pi_u^{-2}$
$a^1\Delta_g$	32.50	33.4	$1\pi_u^{-2}$
$b^1\Sigma_g^+$	(33.25)	33.8	$1\pi_u^{-2}$
$A^3\Pi_u$		37.5	$3\sigma_g^{-1}1\pi_u^{-1}$
$c^1\Pi_u$		38?	$3\sigma_g^{-1}1\pi_u^{-1}$
$^3\Pi_g$ & $^1\Pi_g$?		40	$2\sigma_u^{-1}1\pi_u^{-1}$

References

[1] J.E. Reutt et al., *J. Chem. Phys.*, **84** (1986) 3022.
[2] M. Perić, B. Ostojić and B. Engels, *J. Chem. Phys.*, **109** (1998) 3086.
[3] R. Thissen et al., *J. Chem. Phys.*, **99** (1993) 6590.
[4] Y.H. Jiang et al., *Phys. Rev. Lett.*, **105** (2010) 263002.
[5] A. Kivimaki et al., *J. Phys. B: At. Mol. Opt. Phys.*, **30** (1997) 4278.
[6] J.H.D. Eland et al., *Philos. Trans. R. Soc. London, Ser. A*, **324** (1988) 247.
[7] S.R. Andrews, F.M. Harris and D.E. Parry, *Chem. Phys.*, **166** (1992) 69.
[8] O. Furuhashi et al., *Chem. Phys. Lett.*, **342** (2001) 625.
[9] R.J. Hall et al., *Chem. Phys.*, **187** (1994) 125.
[10] T. Kinugawa et al., *J. Electron Spectrosc. Related Phenom.*, **141** (2004) 143.
[11] B. Gaire et al., *Phys. Rev. A*, **89** (2014) 013403.

4.3 Formaldehyde (methanal), H_2CO

Formaldehyde is planar and has the same number of valence electrons as O_2, but because the orbitals, including the outermost one, are not degenerate the ground state is a spin singlet. The electron configuration can be written in C_{2v} symmetry as

$$O1s^2C1s^2O2s^2C2s^21b_2{}^25a_1{}^21b_1{}^22b_2{}^2............{}^1A_1.$$

In the photoelectron spectrum [1] there are four bands for ionisation out of the outer orbitals at 10.9, 14.5, 16.0 and 16.6 eV. The C2s-based states appear near 21 eV and N2s ionisation produces a band at 34.5 eV [3]. The outermost orbital is relatively non-bonding and is located mainly on the O atom. The first study of the double ionisation spectrum was an examination of its CK and OK Auger spectra [2], which located nine states but gave no attributions. The Auger spectra were theoretically interpreted by Minelli et al. [4], who gave some band assignments. A combined TOF-PEPECO and theoretical study has since been able to characterise the low-energy double ionisation spectrum in greater detail [3], partly because the lifetime broadening which limits resolution in the Auger spectra is absent.

Five bands are seen in the TOF-PEPECO spectrum shown in Fig. 4.3.1. The vertical ionisation energies to the five states are listed below. Four states which can be reached

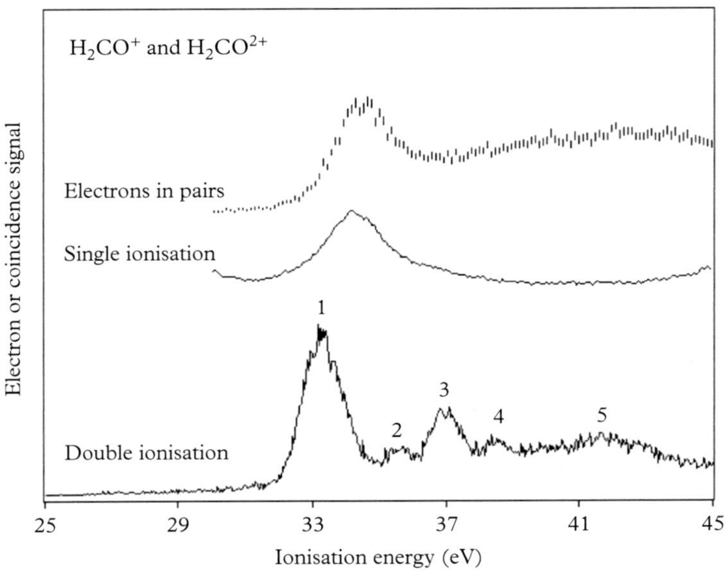

Figure 4.3.1 *The TOF-PEPECO spectrum of formaldehyde (lowest curve), together with part of the single photoionisation spectrum and a spectrum of electrons detected in pairs produced by double photoionisation. All the spectra were taken at 48.37 eV using the same apparatus.*

by removing at least one electron from $2b_2$ have calculated energies matching the first four bands in the spectrum very well. Such simple correspondence may be misleading, however, because combinations of orbital energies suggest, and the calculations confirm, that some of the observed bands have contributions from several states. The band with its peak at 42 eV may come from a higher-order process where one electron is promoted to a normally unoccupied orbital, but the band energy also matches the energy expected if one electron is removed from the inner valence $4a_1$ (C2s) orbital and one from an outer valence orbital.

The spectra in Fig. 4.3.1 demonstrate an important mechanism of double photoionisation, first recognised here [3]. The inner valence band at 34.5 eV in the single ionisation spectrum, whose intensity comes mainly from O2s ionisation, lies above the molecular double ionisation threshold. This makes an *inner valence Auger effect* possible, where single ionisation to states within this band is followed by autoionisation. It strongly enhances the intensity of the first band in the TOF-PEPECO spectrum of formaldehyde. The same process must be expected to play a part in the double ionisation of all molecules with inner valence bands above the double ionisation thresholds. This will be most important for small molecules containing O or N atoms and for large molecules such as polynuclear aromatic hydrocarbons, where the C2s-based states are also above the double ionisation limit.

Double ionisation of H_2CO

Band	Peak (eV)	Theory (eV)	State	Ionisation
1	33.1	33*	X^1A_1	$2b_2^{-2}$
2	35.5	35.6	a^3A_2	$1b_1^{-1}2b_2^{-1}$
3	37.0	36.6	A^1A_2	$1b_1^{-1}2b_2^{-1}$ and c^3B_2 from $4a_1^{-1}2b_2^{-1}$
4	38.6	38.4	B^1B_2	$4a_1^{-1}2b_2^{-1}$
5	42			

* Fixed as standard. More detail is given in ref. [3].

References

[1] P.R. Keller et al., *Chem. Phys.*, **90** (1984) 147.
[2] N. Correia et al., *J. Chem. Phys.*, **95** (1991) 5187.
[3] M. Hochlaf and J.H.D. Eland, *J. Chem. Phys.*, **123** (2005) 164314.
[4] D. Minelli et al., *J. Chem. Phys.*, **99** (1993) 6688.

4.4 Cyanogen, C_2N_2

At equilibrium, the cyanogen molecule is linear and symmetric as NCCN. We can write its electron configuration as

$$N1s^4C1s^4N2s^4C2s^4\pi_u{}^4\sigma_u{}^2\sigma_g{}^2\pi_g{}^4 \ldots\ldots\ldots\ldots\ldots\ldots {}^1\Sigma_g{}^+.$$

In the photoelectron spectrum taken with 21.22 eV photons [1, 2] the first four bands showing ionisation from each of the outer orbitals are at about 13.5, 14.6, 14.9 and 15.7 eV. In a single ionisation spectrum taken in our laboratory using 40.81 eV light, C2s ionisation appears as a double-peaked band centred at 23 eV. Double photoionisation of this molecule had hardly been examined at all before the first study using TOF-PEPECO and electron-ion coincidence techniques together with high-level theory [3]. The binding energies of the outer orbitals are so close together that a large number of dication states must be expected to lie within a relatively small energy range, and this expectation is borne out by the theoretical calculations [3]. A further complication revealed by the calculations is that linear $NCCN^{2+}$ is not the lowest energy isomer of the dication, as linear $CNCN^{2+}$ is more stable by about 1.5 eV. For internal energies greater than about 2 eV, a rearrangement to the lowest energy isomer is calculated to occur rapidly. A substantial fraction (17%) of dissociation to ion pairs after 40.81 eV photoionisation goes through this rearrangement to the products $C^+ + CN_2{}^+$.

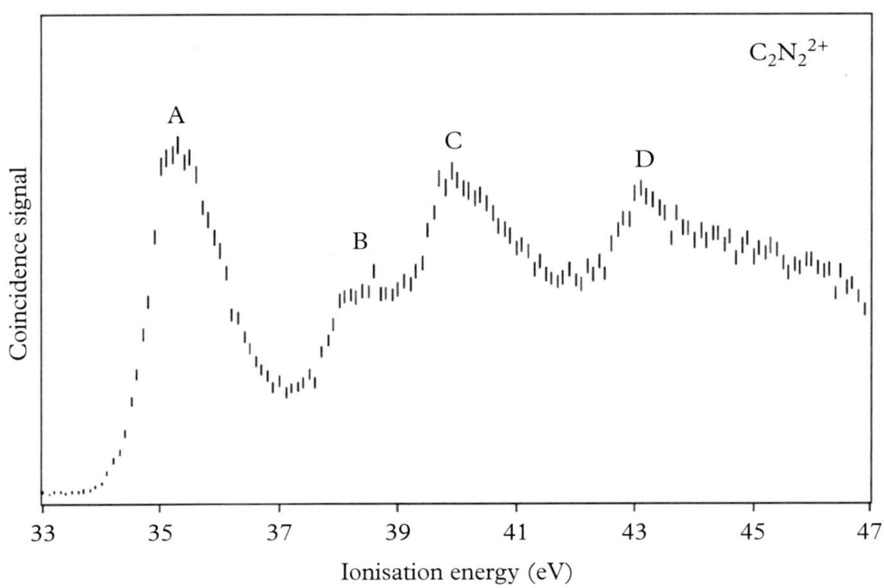

Figure 4.4.1 *TOF-PEPECO spectrum of cyanogen at 48.37 eV photon energy. A spectrum taken at 40.81 eV contains no extra features. Bands A–D are discussed in the text.*

Even removal of a single valence electron from cyanogen causes structural changes producing significant vibrational progressions in the photoelectron spectrum, so it is no surprise that for the already congested electronic states of the dication we cannot resolve any vibrational structure. The TOF-PEPECO spectrum in Fig. 4.4.1 contains four broad bands centred at 35.3, 38.4, 39.9 and 43.1 eV, identified in the figure as A, B, C and D. The detailed calculations and also the sums of orbital energies agree that band A arises from π_g^{-2}, $\sigma_g^{-1}\pi_g^{-1}$ and related ionisations but even here some states have strongly mixed configurations. The states forming band B are of even more mixed configurations, but all seem to involve π_u^{-1} ionisation. Band C has several contributors, among which configurations with π_u^{-2} ionisation are prominent. The highest energy band, D, may arise from configurations where one valence electron has been promoted to a normally unoccupied orbital, but the energy suggests that $C2s^{-1}$ + valence^{-1} ionisation may be a major contributor.

References

[1] L.S. Cederbaum, *Chem. Phys.*, **10** (1975) 459.
[2] M. Hochlaf et al., *J. Chem. Phys.*, **123** (2005) 144302.
[3] M. Hochlaf and J.H.D. Eland, *Phys. Chem. Chem. Phys.*, **10** (2008) 5394.

4.5 Phosphorus trichloride, PCl_3

The PCl_3 molecule has the same pyramidal structure as ammonia and the outermost orbital, $11a_1$, might be thought of as a simple phosphorus 3p lone pair, but *ab initio* calculations [1, 2] show that in fact it has about 45% Cl3p character. The valence orbital configuration of the molecule is

$$9a_1{}^2 8e^4 10a_1{}^2 9e^4 10e^4 2a_2{}^2 11a_1{}^2 \ldots\ldots\ldots\ldots\ldots\ldots {}^1A_1.$$

The next three orbitals after $11a_1$ in order of increasing binding energy have almost pure Cl3p character and the deeper ones have both chlorine and phosphorus atomic orbital constituents. The photoelectron spectrum [3, 4] shows that the $11a_1$ ionisation band at 10.5 eV binding energy is separate, $2a_2$ and 10e overlap at 11.7 and 12 eV, 9e is at 13 eV and $10a_1$ is at 14.2 eV. Some of these binding energies may be useful in interpretation of the double ionisation spectrum, Fig. 4.5.1, as there are two different interpretations of the Auger spectra [5, 6]. Platania et al. [5] interpret their well-resolved Auger spectra to locate the lowest vertical double ionisation at 33 eV using transitions from both the P2p and the Cl2p vacancy states. Larkins et al. [6] on the other hand locate the lowest energy double ionisation band at 29.3 or 28.8 eV from their slightly less well-resolved P2p Auger spectrum.

In Fig. 4.5.1, the first band has its onset at 27.3 eV and peak at 28.2 eV, and so does not exactly match either analysis of the Auger spectra. We interpret the first band as

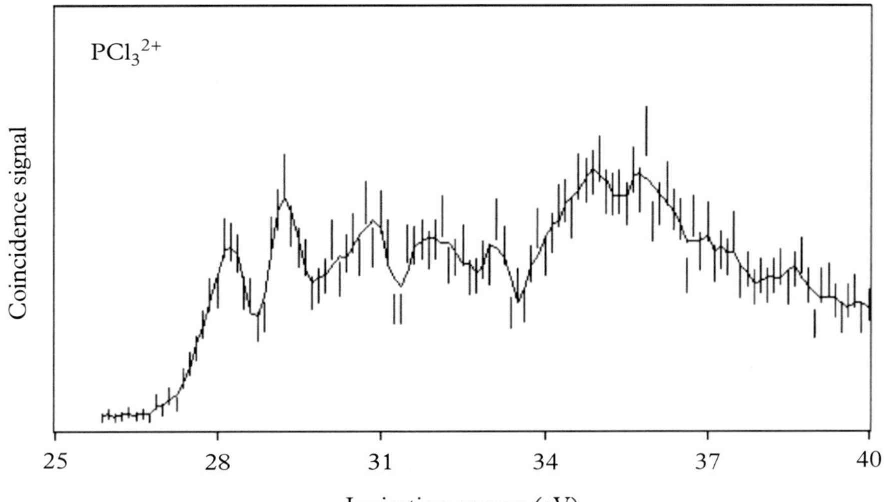

Figure 4.5.1 *TOF-PEPECO spectrum of PCl_3 at 40.81 eV photon energy. A line is drawn following smoothed data to guide the eye between the rather sparse points, shown as error bars.*

the $^1A_1$ state derived by removal of two electrons from the $11a_1$ orbital, and the second band peaking at 29.2 eV as the ^3E state from $10e^{-1}11a_1^{-1}$. The 1 eV spacing between the two bands is well in line with expectation from the difference in the 10e and $11a_1$ binding energies, and also agrees with the 0.8 eV spacing calculated *ab initio* by Larkins et al. [6]. For the higher bands, both combinations of orbital binding energies and the explicit calculations of the P and Cl Auger spectra [6] indicate the existence of many, probably overlapping states. The intensity distribution is sure to be very different in the double photoionisation spectrum as compared with either of the L-shell (2p) Auger spectra. By analogy with the TOF-PEPECO spectra of other molecules containing at least two chlorine atoms (Sections 4.8–4.10), we suggest that the features between 31.5 and 33.2 eV probably represent ionisation of two electrons from a single Cl atom.

References

[1] I.H. Hillier and V.R. Saunders, *J. Chem. Soc., Dalton Trans.*, (1972) 21.
[2] M. Xin et al., *J. Electron Spectrosc. Related Phenom.*, **33** (1984) 93.
[3] A.W. Potts et al., *Philos. Trans. R. Soc. London, Ser. A*, **268** (1970) 59.
[4] D. Bettridge et al., *Anal. Chem.*, **44** (1972) 2005.
[5] P. Platania et al., *Chem. Phys.*, **163** (1992) 381.
[6] F. Larkins et al., *J. Phys. B: At. Mol. Opt. Phys.*, **26** (1993) 1479.

4.6 Methane, CH_4

The tetrahedral methane molecule has a simple electron configuration:

$$1s^2 2a_1{}^2 1t_2{}^6 \ldots\ldots\ldots\ldots\ldots\ldots\ldots\ldots {}^1A_1.$$

Its photoelectron spectrum [1, 2] contains two main bands for single ionisation from the outer orbitals; the band from $1t_2{}^{-1}$ ionisation is split into three components by Jahn–Teller distortion, so the lowest energy state of the $CH_4{}^+$ ion does not retain the T_d geometry of its parent. The resulting band has a complex and largely unresolved vibrational structure centred at 14.5 eV. The $^2A_1$ band at 23 eV from removal of a C2s ($2a_1$) electron is resolved into a vibrational progression in the C—H stretching vibration.

The double ionisation spectrum of methane has been studied by multiple techniques and the early experiments and theoretical work were brought together in a comprehensive paper by Griffiths et al. [3]. When two electrons are removed from the $1t_2$ orbital, four states can be produced with the symmetries $^3T_1$, ^1E, $^1T_2$ and $^1A_1$ in T_d geometry, but it is very unlikely that they stay tetrahedral for long. No stable or metastable $CH_4{}^{2+}$ ions are found in mass spectra where ionisation is direct from neutral methane, but

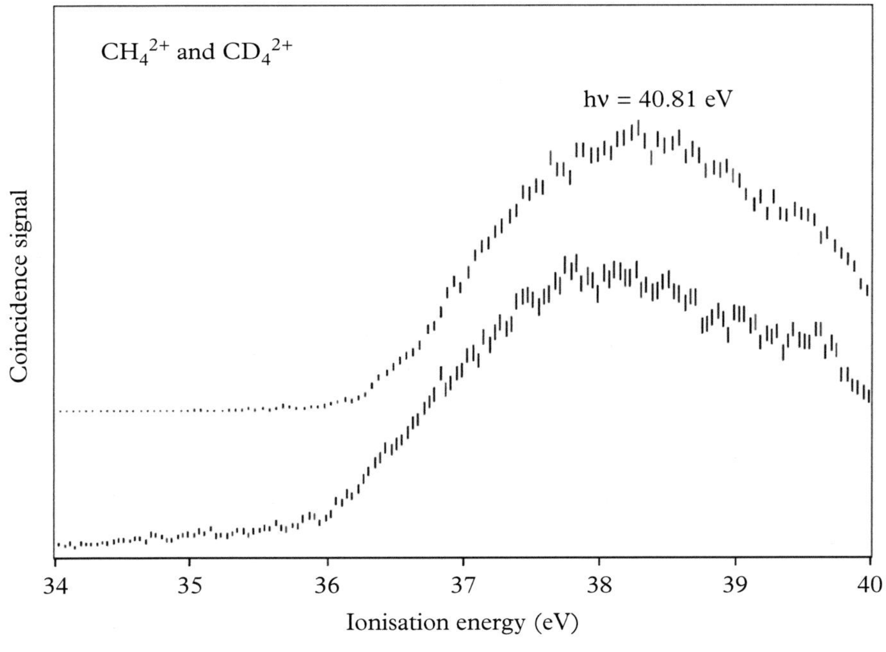

Figure 4.6.1 *TOF-PEPECO spectra of methane (lower curve) and methane-d$_4$ (upper curve) at 40.81 eV photon energy showing the shift in onset of the main band on deuteration.*

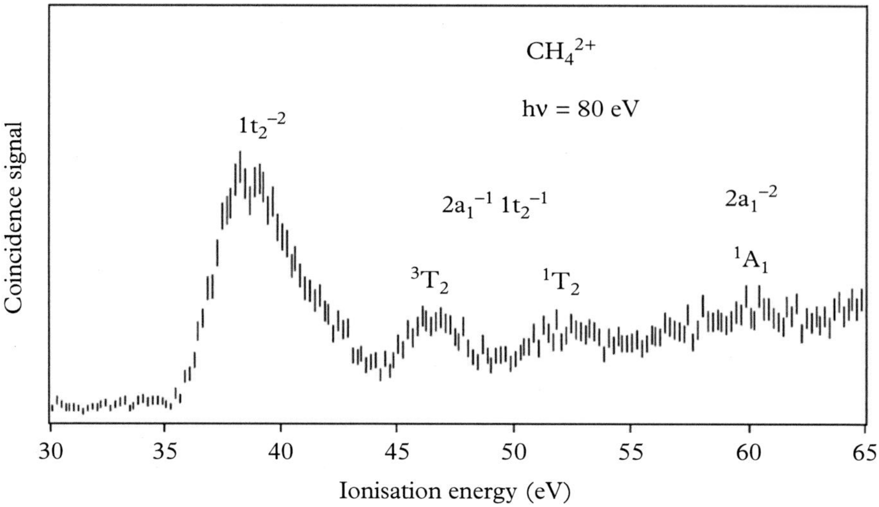

Figure 4.6.2 *TOF-PEPECO spectrum of methane at 80 eV photon energy.*

metastable square-planar CH_4^{2+} ions are formed when singly ionised CH_4^+ is ionised further [4, 5].

The TOF-PEPECO spectra of methane taken with laboratory He radiation [6], Fig. 4.6.1, add only a little to the knowledge already gained from the double charge transfer (DCT) [3] and Auger spectroscopies [7], and from theory [3]. The main band with peak near 38 eV certainly represents ionisation to the $^3T_1$ state, possibly with contributions from ^1E and $^1T_2$ on the high-energy side. The main onset is at about 36 eV, but some real intensity from indirect double ionisation extends down to 35.1 eV. Surprisingly, both for CH_4 and CD_4 the low-energy edges of the $^3T_1$ bands show vestigial but reproducible vibrational structure, with intervals consistent with C—H or C—D stretching vibration [6]. There is a significant shift of the main onset on deuteration due to contraction of the Franck–Condon zone, which makes the first vertically accessible levels of the doubly ionised ion higher up on the repulsive surface. The CH_4^{2+} state energies listed below are condensed estimates from the work of Griffiths et al. [3].

When methane is doubly photoionised at a photon energy above 45 eV, the band centred at about 38 eV, which contains all four states arising from $1t_2^{-2}$ ionisation, is broadened on the high-energy side as seen in Fig. 4.6.2. Three higher-energy states, whose formation involves ejection of one or two electrons from the inner valence $2a_1$ orbital, are visible as broad bands. At this higher photon energy, no indirect ionisation populating dissociative states below 36 eV is apparent. The tendency for the relative importance of indirect double ionisation processes to diminish as photon energy increases is quite general, because cross-sections for direct double ionisation increase with photon energy, while those for formation of singly ionised intermediate states remain constant.

Vertical ionisation energies (eV) to states of CH_4^{2+}

State	Energy	Orbitals
$^3T_1$	38	$1t_2^{-2}$
^1E	39	$1t_2^{-2}$
$^1T_2$	40	$1t_2^{-2}$
$^1A_1$	43	$1t_2^{-2}$
$^3T_2$	47	$2a_1^{-1}1t_2^{-1}$
$^1T_2$	52	$2a_1^{-1}1t_2^{-1}$
$^1A_1$	61	$2a_1^{-2}$

References

[1] G. Bieri and L. Åsbrink, *J. Electron Spectrosc. Related Phenom.*, **20** (1980)149.
[2] M. Carlsson Göthe et al., *J. Chem. Phys.*, **94** (1990) 2536.
[3] W.J. Griffiths et al., *Rapid Commun. Mass Spectrom.*, **6** (1992) 438.
[4] T. Ast et al., *Chem. Phys. Lett.*, **78** (1981) 439.
[5] P.E.M. Siegbahn, *Chem. Phys.*, **66** (1982) 443.
[6] J.H.D. Eland, *Chem. Phys.*, **323** (2006) 391.
[7] I.B. Ortenburger and P.S. Bagus, *Phys. Rev. A*, **11** (1975) 1501.

4.7 Methyl halides (monohalomethanes), CH_3F, CH_3Cl, CH_3I

The methyl halides CH_3F, CH_3Cl, CH_3Br and CH_3I all have the same spatial structure and symmetry, belonging to the C_{3v} point group. Their electronic structure in the outer shells also has the same orbital order, in which the highest occupied molecular orbital (HOMO) has more and more lone-pair character and shows more spin–orbit splitting as we go to the heavier halogens. The orbital makeup of CH_3F is slightly different from that of the others because F2p has higher binding energy than C2p, whereas in the other molecules the halogen p-orbitals are least strongly bound. The configuration of the valence shell and inner valence orbitals outside the deep inner shells (C_{3v} symmetry) is

$$\mathrm{Xns^2C2s^21e^4a_1{}^22e^4 \ldots\ldots\ldots\ldots\ldots\ldots {}^1A_1}.$$

In CH_3F the outermost (2e) orbital has more C2p than halogen 2p character but the opposite is true in the heavier halides. We start with wide-range double ionisation spectra of methyl fluoride at two photon energies, in Fig. 4.7.1.

Methyl fluoride has no stable doubly charged states; no doubly charged parent ions are found in its mass spectrum; and there is no resolved vibration structure in the double ionisation spectrum. All the dicationic states initially formed by photoionisation dissociate rapidly to such products as $CH_3{}^+ + F^+$ (thermodynamic onset 32.1 eV) or $CH_2F^+ + H^+$ (onset 27.0 eV). The spectra illustrate two phenomena which are found in many spectra. Firstly, the relative intensities of the band structures (but not their energies) change systematically as the photon energy is changed. Bands at higher ionisation energy become more intense at higher photon energy. Secondly, there are clear features above the onset of triple ionisation, which is located near 70 eV according to our own e–e–e triple coincidence measurements. States at energies above this limit can dissociate competitively to either doubly or triply charged fragments, and may do so by single-step or multi-step pathways.

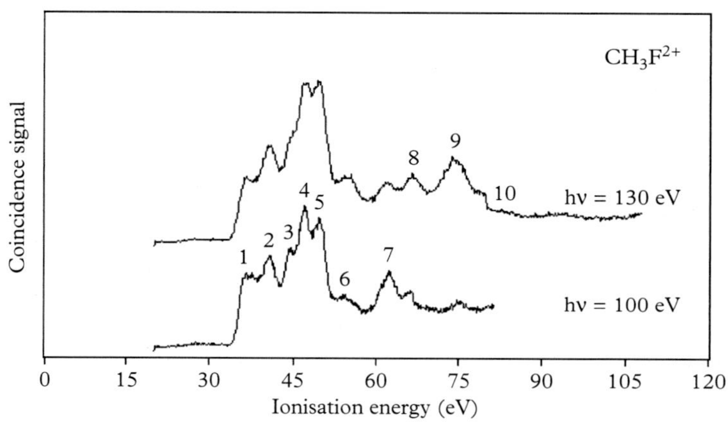

Figure 4.7.1 *Double photoionisation TOF-PEPECO spectra of CH_3F at the two energies indicated. Spectra taken in the laboratory at higher resolution using HeII light show no additional structure. Energies of the numbered bands are listed later.*

Many features with some resemblance to the bands in Fig. 4.7.1 are seen in the carbon and particularly the fluorine Auger spectra, which are presented and analysed by Liegener [1]. Unfortunately, very few peaks occur at the same energies in the two Auger spectra, and none match the peaks in Fig. 4.7.1 closely enough for clear identification. According to the theoretical calculations, all the states have mixed configurations, but single configurations are dominant at the lowest and also at the highest energies. In view of this we use the electron configuration of the neutral CH_3F molecule to discuss those states. The binding energies of the three outermost orbitals taken from the photoelectron spectrum [2] are 12.6, 17 and 23 eV and the lowest dication states, located around 36 eV, are attributed to the loss of two electrons from the outer 2e orbital. Using the very crude model first introduced for HBr, this implies a Coulomb repulsion energy of about 11 eV and hence an intercharge distance of about 1.3 Å. The configuration e^2 (in C_{3v} symmetry) formally produces the states $^3A_2$, ^1E and $^1A_1$. The two singlet states are calculated [1] to be at 35.8 and 36.3 eV; the triplet will also be very close and probably lowest in energy. The peak at 40.7 eV may arise from an $a_1^{-1}2e^{-1}$ ionisation (calculated 39.7 eV [1]) or from $1e^{-1}2e^{-1}$ (calculated 41.4 eV) or from both. The states from 45 eV upwards have heavily mixed configurations, with the exception of one formed by almost pure F2s electron pair loss at 98 eV [3], faintly visible in Fig. 4.7.1. For examples of two of the heavier methyl halides, TOF-PEPECO spectra taken with HeII light are shown together in Fig. 4.7.2.

The spectrum of methyl chloride shows three main bands with peaks at roughly 31.6, 35 and 38 eV; the first band shows signs of three unresolved components with individual

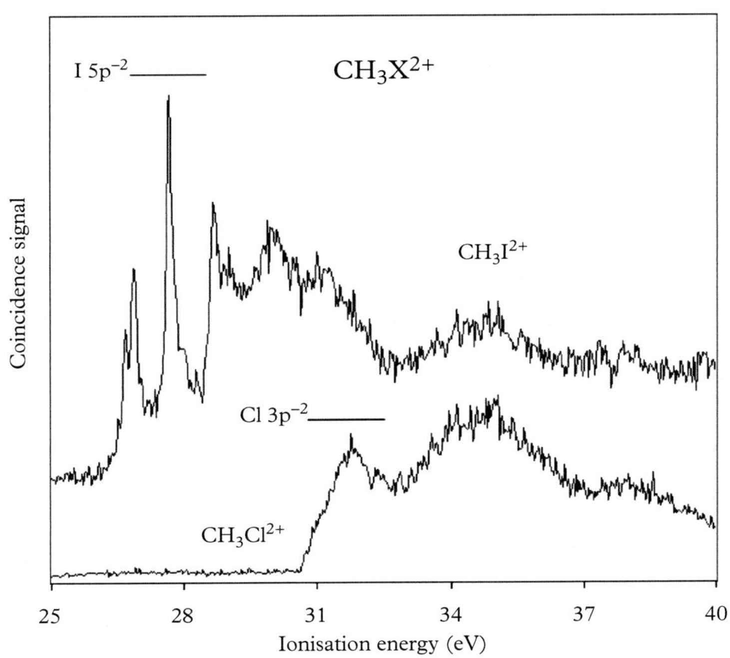

Figure 4.7.2 *TOF-PEPECO spectra of methyl chloride and methyl iodide at 40.81 eV photon energy.*

peaks at 31.0, 31.7 and 32.3 eV and first onset at 30.5 eV. In methyl iodide, the lowest energy part contains three rather sharp bands, each of which shows further splitting when measured at higher resolution [4]. The same bands could be recognised in the iodine 4d Auger spectrum [5]. The considerable difference between the spectra of the two halides comes from a difference in electronegativity of the atoms, together with the strong spin-orbit interaction in iodine which makes the spectrum of CH_3I resemble that of HI (Fig. 2.2.1 and ref. [4]). Whereas the p_{xy} electrons of the Cl atom perpendicular to the C—Cl bond interact quite strongly with the p_{xy} electrons of carbon in CH_3Cl creating real molecular orbitals, those of iodine interact much less strongly because of the disparity in size and binding energy, and remain highly localised. The lowest energy double ionisation process for both molecules is removal of two electrons from p_{xy} orbitals located mainly on the halogen, resulting in the three states $^3\Sigma^-$, $^1\Delta$ and $^1\Sigma^+$ of a linear molecule reduced to $^3A_2$, ^1E and $^1A_1$ symmetry, and perturbed to a variable extent by bonding and the departure from cylindrical symmetry. The first band in the double ionisation spectrum of methyl iodide clearly shows spin-orbit splitting of the triplet term into what would have been the $\Omega = 0$ and $\Omega = \pm 1$ levels in cylindrical symmetry. This spectrum was examined experimentally in some detail at photon energies from 32.7 eV to 51.0 eV and has been interpreted theoretically by Pernpointner et al. [6] using the Green's function methodology including full treatment of relativistic effects. The computed spectra extend up to 100 eV ionisation energy and thus include double ionisation involving the I4d orbitals, but the Auger effect which would inevitably follow single ionisation from I4d was not explicitly considered.

Energies (eV) of the sharp peaks in the spectrum of CH_3I^{2+} with symmetries in C_{3v} followed by the equivalents for a linear molecule in parentheses

$^3A_2$('$^3\Sigma_0^-$')	26.7
$^3A_2$ ('$^3\Sigma_{\pm1}^-$')	26.8
^1E ('$^1\Delta$')	27.7
$^1A_1$ ('$^1\Sigma_0^+$')	28.7

Ionisation energies (eV) of the numbered band peaks for CH_3F^{2+} (Fig. 4.7.1)
1: 36.4 2: 40.7 3: 44.2 4: 46.4 5: 49.3 6: 54.2 7: 62.1 8: 65.7 9: 73.6 10: 82.2.

References

[1] C.-M. Liegener, *Chem. Phys. Lett.*, **151** (1988) 83.

[2] T.N. Olney et al., *J. Electron Spectrosc. Related Phenom.*, **66** (1993) 83.

[3] R. Feifel et al., *Phys. Rev. Lett.*, **116** (2016) 073001.

[4] A. Pilcher-Clayton and J.H.D. Eland, *J. Electron Spectrosc. Related Phenom.*, **142** (2005) 3132.

[5] D.M.P. Holland et al., *Chem. Phys.*, **326** (2006) 535.

[6] M. Pernpointner et al., *Chem. Phys.*, **407** (2012) 39.

4.8 Methylene halides (dihalomethanes), CH_2Cl_2, CH_2Br_2, CH_2I_2

The outer orbitals of the dihalomethanes from methylene chloride to methylene iodide are generally thought of as non-bonding halogen lone-pair orbitals, separate from the bonding orbitals. The ionisation energies of the four non-bonding orbitals in each compound are quite close together, promising complex overlapping states in the double ionisation spectra. The outer orbital configuration common to all in C_{2v} symmetry, with a dotted space separating the more tightly bound and mainly bonding orbitals from less tightly bound and mainly non-bonding ones is

$$\text{C1s}\,(a_1)^2\,1b_1{}^2 1a_1{}^2 1b_2{}^2 \ldots 2a_1{}^2 2b_1{}^2 1a_2{}^2 2b_2{}^2 \ldots\ldots\ldots\ldots\ldots\ldots {}^1A_1.$$

The photoelectron spectra of all the dihalomethanes were reported by Potts et al. [1] and give the ionisation (orbital) energies, but these are of limited use in interpretation of the TOF-PEPECO spectra shown in Fig. 4.8.1.

To interpret the double ionisation spectra we can think of the eight electrons in non-bonding orbitals, whose pairwise removal will give rise to the lowest doubly charged ion states, as roughly equivalent to the eight electrons in the two degenerate delocalised molecular π-orbitals of the diatomic halogens such as I_2. If the interhalogen bonds were progressively lengthened, the splitting between π_g and π_u would diminish and the p_{xy} atomic orbitals on the two halogens would become independent and equivalent. The main distinction in double ionisation would then be whether two electrons are removed from a single halogen atom or one is removed from each. A TOF-PEPECO study of the double ionisation of series of dihalogen and other difunctional molecules [2] clearly confirms the applicability of this model, as the distance between the separated charges determines the lowest double ionisation energy. If one electron is removed from each of two independent halogen atoms, each one will have two possible states $^2\Pi_{3/2}$ and $^2\Pi_{1/2}$ with equal statistical weight in its local cylindrical symmetry. The double ionisation spectrum will then contain three peaks, equally spaced at the same spacing as in CH_3X^+, with

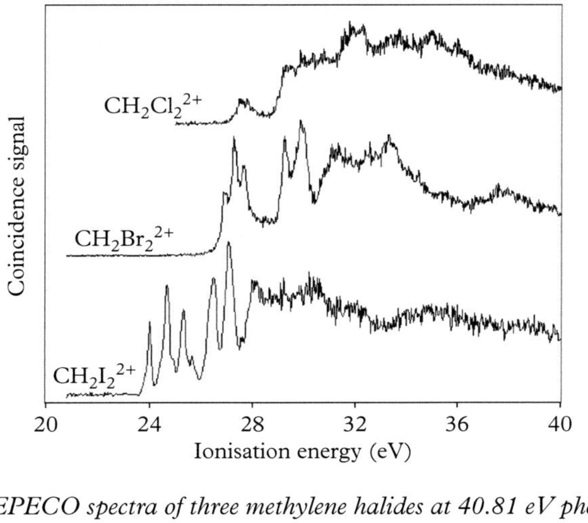

Figure 4.8.1 *TOF-PEPECO spectra of three methylene halides at 40.81 eV photon energy.*

1:2:1 relative intensities (3/2,3/2; 3/2,1/2 + 1/2,3/2; 1/2,1/2). This is exactly what we see as the lowest energy bands in the double ionisation spectra of CH_2I_2 and CH_2Br_2. For the diiodide bands, the spacings are 0.7 and 0.6 eV (0.66 eV in CH_3I^+); for the dibromide, the spacings are 0.4 and 0.3 eV (0.32 eV in CH_3Br^+). The spin-orbit splitting for chlorine is too small to be seen. A further prediction is that if two electrons are taken from the same halogen atom, the usual three-peak pattern should be seen at about the same energy as in the lowest double ionisation of HX or CH_3X. This is well borne out in the spectrum of the diiodide (peaks at 26.5, 27.1 and 28.1 eV, cf. 26.7, 27.7 and 28.7 eV in CH_3I^{2+}) and similarly for the dibromide. The authors have not found any Auger or other studies of the double ionisation spectra of these compounds, but there has been considerable activity in examination of their fragmentation after double ionisation [3–5].

Band centres of features in the spectrum of $CH_2Cl_2^{2+}$

Energy (eV)	Description
27.6	$3p^{-1}$ on separate Cl atoms, outermost orbital
29.4	2 × $Cl3p^{-1}$, probably mixed orbitals
32.0	$3p^{-2}$ on a single Cl atom (cf. 32 eV in CH_3Cl, Fig. 4.7.2)

Band peaks in the spectrum of $CH_2Br_2^{2+}$

Energy (eV)	Description
26.9, 27.3, 27.6	$Br4p^{-1}$ twice: (2 $\times ^2P_{3/2}$, $^2P_{3/2} + ^2P_{1/2}$, 2 $\times ^2P_{1/2}$)
29.2, 29.9, 30.8	Single $Br4p^{-2}$
33.4	Broad bump

Band peaks in the spectrum of $CH_2I_2^{2+}$

Energy (eV)	Description
24.0, 24.6, 25.3	$I5p^{-1}$ twice: (2 $\times ^2P_{3/2}$, $^2P_{3/2} + ^2P_{1/2}$, 2 $\times ^2P_{1/2}$)
26.4, 27.0, 28.1	Single $I5p^{-2}$ (cf. 26.7, 27.7, 28.7 eV in CH_3I)
30.4	Broad bump

References

[1] A.W. Potts et al., *Philos. Trans. R. Soc. London, Ser. A*, **268** (1970) 59.
[2] R.D. Molloy et al., *Chem. Phys.*, **335** (2007) 49.
[3] A.F. Lago et al., *J. Phys. Chem. A*, **110** (2006) 13717.
[4] K.F. Alcantra et al., *J. Phys. Chem. A*, **119** (2015) 8822.
[5] K.F. Alcantra et al., *Chem. Phys.*, **429** (2014) 1.

4.9 Carbon tetrahalides, CF_4, CCl_4, CBr_4

Because of their tetrahedral symmetry, the carbon tetrahalides have a simple valence electron configuration despite the number of heavy atoms:

$$[\text{Core}]\ [\text{IV}]\ a_1{}^2 t_2{}^6 e^4 t_2{}^6 t_1{}^6 \ldots\ldots\ldots\ldots\ldots\ldots\ldots\ {}^1A_1.$$

In the photoelectron spectrum of CF_4 [1, 2] the ionisation energies of the two outermost orbitals are centred at 16.32 and 17.4 eV, while in that of CCl_4 [3] the first two bands are at 11.7 and 12.6 eV. In $CBr_4{}^+$, the strong spin-orbit interaction combines with Jahn–Teller effects to split the degeneracy of the states formed by ionisation from the t_1 and t_2 orbitals, producing bands with peaks at 10.4, 10.8, 11.1, 11.3 and 11.7 eV [4]. Because the outer orbitals are degenerate, more than one electronic state can be generated by taking two electrons out of each, or by taking electrons from different orbitals. As a result of this and of the close orbital energy spacing, there are many accessible states of the dications and the double ionisation spectra are highly congested. No stable $CX_4{}^+$ ions exist for CF_4 or CCl_4 [5] and we do not expect doubly charged parent ions to be stable or metastable for any of the tetrahalides.

The double ionisation of CF_4 had been investigated by Auger spectroscopy [6, 7], DCT spectroscopy [7] and TPEsCO [8] before TOF-PEPECO [9] was applied to it.

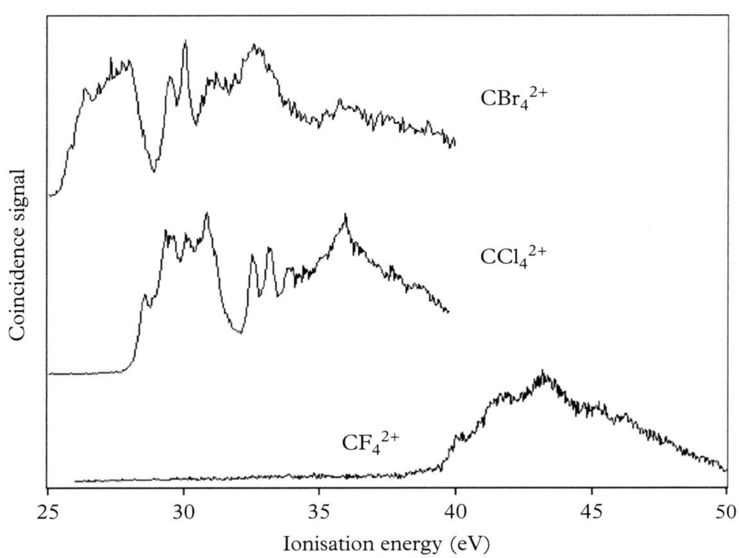

Figure 4.9.1 *TOF-PEPECO spectra of CF_4 at $h\nu$ = 51.02 eV and of CCl_4 and CBr_4 at $h\nu$ = 40.81 eV.*

Each of the experimental approaches was accompanied by theoretical efforts to understand the spectra, and the Green's function method was successful in reproducing the form of the TOF-PEPECO spectrum shown in Fig. 4.9.1. While the Auger spectra are dominated by singlet states, the double charge transfer (DCT) technique can select singlet or triplet states by choice of the projectile used, whereas TOF-PEPECO spectra include both singlets and triplets. The DCT experiments and the accompanying theory [7] show that the lowest energy double ionisation band of CF_4^{2+} involves an overlap of several states, both singlet and triplet, derived by removing two electrons from the outermost (t_1) orbital or one from t_1 and one from the outer t_2. The same conclusion can be drawn from the Green's function theoretical analysis of the TOF-PEPECO spectrum [9], where the most useful classification of spectral features was based on the location of the holes. The lowest energy double ionisation bands, comprising several states, arise from removal of one non-bonding electron from each of two fluorine atoms, so minimising Coulomb repulsion.

Investigation of the double ionisation of CCl_4 was also begun by Auger spectroscopy [10] followed by DCT [11, 12], and the Auger spectrum was later reinterpreted theoretically [13]. The most detailed analysis in terms of individual configurations and states was done together with a high resolution DCT study on CCl_4 by Grant et al. [12] who used semi-empirical MSXα calculations. This work shows very clearly that all the distinguishable bands to be seen in the TOF-PEPECO and other spectra contain contributions from several states and several dominant configurations, but we believe that they can also be usefully described in terms of the location of the final charges. The same is true of the double ionisation of CBr_4, which does not seem to have been examined before.

There is a very strong family resemblance between the spectra of CCl_4^{2+} and CBr_4^{2+} in Fig. 4.9.1, for which we believe that a quite simple interpretation is possible. The lowest energy bands must arise from removal of two non-bonding electrons of the t_1 orbital, *leaving the positive charges on two different halogen atoms*. Separation of the charges is demonstrated by the very small Coulomb repulsion energy deduced from the ionisation energies of the first peaks, relative to twice the orbital energies. For CCl_4, the first double ionisation peak is at 28.55 eV, which is only 5.1 eV more than twice the outer orbital energy of 11.7 eV. The simplistic interpretation is that the charges are 2.8 Å apart, which is too far for charges on a single atom. We ascribe the first group of overlapping bands in the spectrum of CCl_4^{2+}, between 28 and 31 eV, to removal of 3p electrons from distinct Cl atoms. The three peaks between 32 and 34 eV are most naturally described as ionisation of two 3p electrons from a single chlorine atom, as the energy agrees with the less resolved but similar bands in the spectra of CH_3Cl^{2+} and $CH_2Cl_2^{2+}$.

The same interpretation applies even more clearly to the spectrum of carbon tetrabromide double ionisation. The three bands between 29 and 31 eV agree closely in position with the similar bands of $CH_2Br_2^{2+}$. The low ionisation energy bands, which we ascribe to removal of electrons from separate Br atoms, are more congested than in the chloride, presumably because of the increased spin–orbit coupling.

Centres of features in the spectrum of CF_4^{2+}

40.1 eV, 41.6 eV, 43.3 eV, 45.2 eV, 46.4 eV

Distinct peaks in the spectrum of CCl_4^{2+}

28.5 eV, 29.5 eV, 30.15 eV, 30.85 eV, 32.55 eV, 33.15 eV, 33.9 eV, 36 eV

Distinct peaks in the spectrum of CBr_4^{2+}

25.85 eV, 26.35 eV, 27.85 eV, 29.5 eV, 30.05 eV, 31.05 eV, 32.6 eV

References

[1] D.R. Lloyd and P.J. Roberts, *J. Electron Spectrosc. Related Phenom.*, 7 (1975) 325.
[2] A.J. Yencha et al., *J. Electron Spectrosc. Related Phenom.*, **70** (1994) 29.
[3] T.A. Carlson et al., *J. Chem. Phys.*, 77 (1982) 5340.
[4] A.W. Potts et al., *Philos. Trans. R. Soc. London A*, **268** (1970) 59.
[5] A.S. Werner, B.P. Tsai and T. Baer, *J. Chem. Phys.*, **60** (1974) 3650.
[6] F.O. Gottfried, L.S. Cederbaum and F. Tarantelli, *J. Chem. Phys.*, **104** (1996) 9754.
[7] W.J. Griffiths et al., *Chem. Phys.*, **173** (1993) 109.
[8] R. Hall et al., *Chem. Phys.*, **187** (1994) 125.
[9] R. Feifel et al., *J. Chem. Phys.*, **125** (2006) 194318.
[10] R.R. Rye and J.E. Houston, *J. Chem. Phys.*, 78 (1983) 4321.
[11] M. Langford et al., *Int. J. Mass Spectrom. Ion Processes*, **112** (1992) 285.
[12] R.P. Grant et al., *Int. J. Mass Spectrom. Ion Processes*, **142** (1995) 117.
[13] E. Fainelli et al., *J. Electron Spectrosc. Related Phenom.*, **119** (2001) 81.

4.10 Germane and germanium halides, GeH_4, $GeCl_4$, $GeBr_4$

Germanium compounds are interesting for laboratory studies of double ionisation because of the convenient binding energy of the 3d orbital, below 40 eV but above the double ionisation thresholds. This means that analysis of data taken with HeII light can yield spectra from ionisation exclusively by the Auger effect, or entirely excluding the Auger effect. Within the Auger effect, the phenomenon known as 'post-collision interaction' (PCI) can be examined in a new way, and this was the main motivation for the original TOF-PEPECO work [1]. In this phenomenon, a relatively low-energy photoelectron initially ejected by photon absorption may be overtaken by a faster Auger electron emitted later. When this happens, the residual positive charge experienced by the photoelectron increases from 1 to 2, while the charge seen by the Auger electron decreases from 2 to 1. The Auger electron gains energy, while the photoelectron loses energy. The theoretical model of this energy exchange process by Kuchiev and Scheinerman [2] could be tested and was verified by examination of the TOF-PEPECO spectrum of $GeCl_4$. Our purpose here is to illustrate some systematic differences between direct double photoionisation and the Auger process.

In Fig. 4.10.1 we see spectra of germane, germanium tetrachloride and tetrabromide, all either from direct double photoionisation or from Auger decay of the $Ge3d_{5/2}$ hole

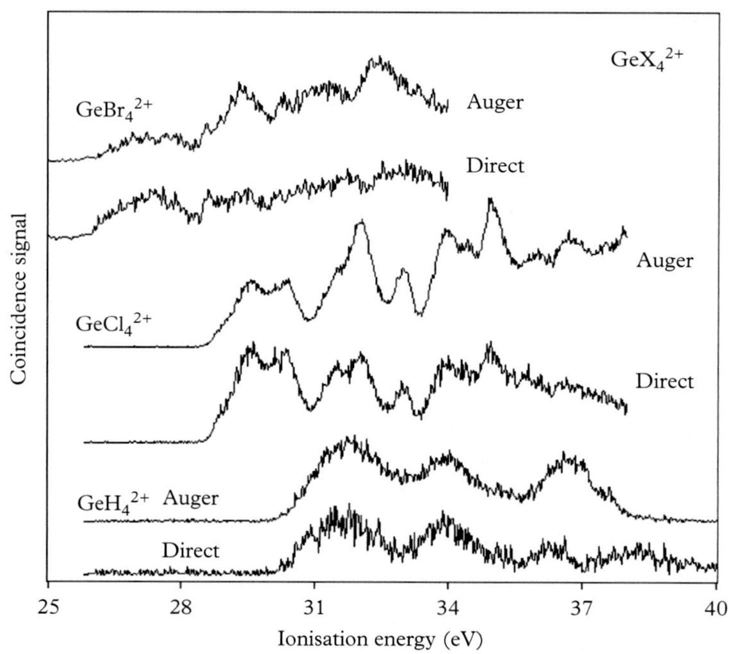

Figure 4.10.1 *TOF-PEPECO spectra of tetravalent Ge compounds taken with 40.81 eV HeII light and extracted from the full data either excluding the Ge3d*$^{-1}$ *states or in coincidence with Ge3d*$_{5/2}^{-1}$ *(marked as Auger).*

state. There is no detectable difference in the spectra when the $Ge3d_{3/2}$ hole state is chosen instead. In the direct spectrum of GeH_4^{2+}, the four broad bands must come from t_2^{-2} ionisation, essentially from the 4p orbitals of Ge itself. Four states, $^3T_1$, ^1E, $^1T_2$ and $^1A_1$, are expected, and we provisionally assign the bands to these states in the order given. Relative intensities of the high-energy bands are probably reduced by exclusion in the analysis of electrons with the same energy as the photolines. In the Auger spectrum, the highest energy part is cut off as no states of higher energy than the photolines can be populated. The major difference between the two spectra is that the third band, assigned as $^1T_2$, is displaced and much stronger in Auger than in direct photoionisation. The reason for this is not clear, but it may relate to the fact that the lowest energy d^{-1} hole state also has $^2T_2$ symmetry. In this case the intermediate d^{-1} hole cannot be treated as an isolated inner shell, as it interacts significantly with the bonding orbitals.

In the germanium halides, the lowest energy single and double ionisation bands represent ionisation from the non-bonding halogen p-orbitals, which have very little overlap with the Ge3d orbital where the hole is created. As a result, these bands are relatively much weaker in the Auger double ionisation spectra of both $GeCl_4$ and $GeBr_4$ than in direct double photoionisation. Bands at higher ionisation energy are uniformly stronger in the Auger spectra.

The general character of some bands in the germanium tetrahalide spectra is evident from comparison with spectra of the dihalomethanes and tetrahalomethanes. The lowest bands come from removal of non-bonding electrons from two different halogen atoms and the next bands come from removal of two from the same halogen atom. The triplet of bands from ionisation at a single Cl atom is particularly clear in the spectrum of $GeCl_4^{2+}$, with relative intensities and spacings close to those in CCl_4^{2+}.

Double ionisation energies for GeH_4 from direct photoionisation

Onset 30.5 eV. Band peaks 31.6, 33.9, 36.3 and 38.3 eV

Double ionisation energies for $GeCl_4$ from direct photoionisation

Onset 28.75 eV. Band peaks 29.6, 30.4, 31.4, 32.1, 33.0, 33.9 and 35.0 eV

Double ionisation energies for $GeBr_4$ from direct photoionisation

Onset 26.1 eV. Band centres 27.2 and 29.0 eV

References

[1] J.H.D. Eland, *Chem. Phys. Lett.*, **409** (2005) 245.
[2] M.Yu. Kuchiev and S.A. Scheinerman, *Sov. Phys. JETP*, **63** (1986) 986.

4.11 Sulphur hexafluoride, SF_6

Although SF_6 has seven atoms, it seems appropriate to include it here as the last centro-symmetric molecule. With so many heavy atoms, it has many orbitals in its electron configuration

$$[\text{Core}]^{22}\,[\text{IV}]^{12}\,5a_{1g}{}^{2}4t_{1u}{}^{6}1t_{2g}{}^{6}3e_{g}{}^{4}1t_{2u}{}^{6}5t_{1u}{}^{6}1t_{1g}{}^{6}\ldots\ldots\ldots\ldots\ldots\,{}^{1}A_{1g}$$

where IV stands for inner valence. The seven valence orbitals produce six distinct bands in the photoelectron spectrum [1] with the first at 15.8 eV and the deepest-lying at 27 eV. Unfortunately, these orbital ionisation energies are of little help in interpreting the double ionisation spectrum, except to warn us that several two-hole states of the doubly charged ion will lie well above the energies of the HeII lines. This expectation is borne out by the Auger spectrum [2, 3], which shows two major bands of intensity, one from about 40 to 65 eV and a second covering about 65 to 80 eV. The first group was interpreted as having two holes in valence orbitals; the second as having one vacancy in the valence and one in the inner valence range. None of the dication states are stable or metastable because no $SF_6{}^{2+}$ ions are observed in any mass spectrum. The TOF-PEPECO spectrum [4] is slightly better resolved than the Auger spectra in the low-energy range.

The TOF-PEPECO spectra taken at 40.81 and 48.37 eV (see Fig. 4.11.1) show only broad and overlapping features. Examination of the single electron spectrum at the same photon energy reveals that there are superexcited singly ionised states covering

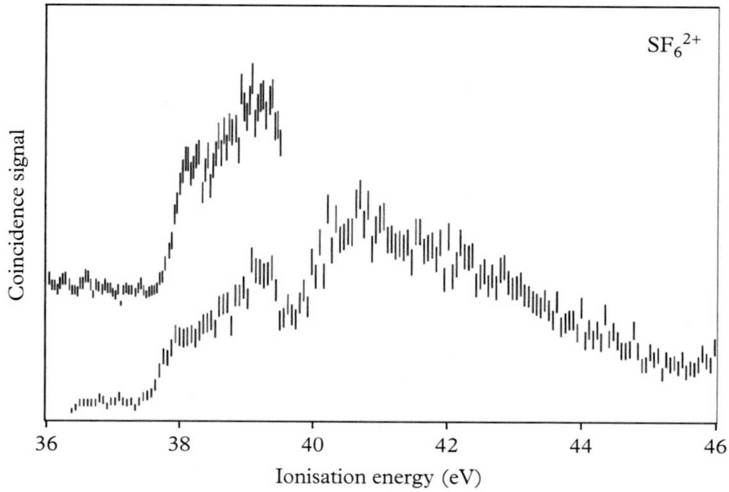

Figure 4.11.1 *TOF-PEPECO spectra of SF_6 at 40.81 eV (upper data set) and 48.37 eV photon energy, shown as error bars because of the low number of counts.*

almost the same energy range as the double ionisation bands, almost certainly undergoing autoionisation and complicating the spectrum. By careful but somewhat arbitrary selection of the electron energy ranges to be included in building up the double ionisation spectrum, the contrast between features could be enhanced but no new features were found [4]. The onset of double ionisation is at 37.6 eV and there are fairly distinct features peaking at 38, 39.2 and 47 eV. The spectrum was interpreted theoretically [4] using a Green's function method where the contributing states were subdivided into four categories: states with two holes localised on a single F atom; states with two holes on different neighbouring F atoms; states with holes on two non-neighbouring ('distal') F atoms; and states with one hole on F and one on S. The lowest energy state of nascent SF_6^{2+}, to which we attribute the onset of the spectrum at 37.7±.2 eV and the incipient peak at 38 eV, is ascribed to removal of two F2p electrons. The calculations suggest that the state involved has about equal contributions from two 2p vacancies on a single atom and two single vacancies on separate atoms. As the calculations predicted several thousand electronic states within the energy range of the measured spectra, it is not possible to give more precise attributions. Previous work on the double ionisation spectrum of this molecule has included Auger spectra [2], DCT spectroscopy [5] and TPEsCO [6]. References to the rather extensive studies devoted to fragmentation of nascent SF_6^{2+} are given in connection with the TOF-PEPECO work [4].

References

[1] D.M.P. Holland et al., *Chem. Phys.*, **192** (1995) 333.
[2] G.M. Bancroft et al., *Phys. Rev. A*, **41** (1990) 3716.
[3] Y. Sato et al., *Chem. Phys. Lett.*, **196** (1991) 475.
[4] R. Feifel et al., *J. Chem. Phys.*, **122** (2005) 144309.
[5] W.J. Griffiths and F.M. Harris, *Int. J. Mass Spectrom. Ion Processes*, **85** (1988) 259.
[6] A.J. Yencha et al., *J. Phys. B; At. Mol. Opt. Spec.*, **33** (2000) 945.

Chapter 5

Conjugated and aromatic molecules

5

Conjugated and aromatic molecules

In the vast majority of conjugated and aromatic molecules, the outermost occupied orbitals are either of π character or are non-bonding lone pairs belonging to heteroatoms. These are the orbitals from which double ionisation gives rise to most of the distinct bands that can be discerned in their spectra. The more tightly held and more strongly bonding σ-electrons are usually much more numerous and close together in binding energy, and give rise to a quasi-continuum of double ionisation bands. As the molecules get bigger, the threshold energy for molecular double ionisation becomes lower, and can become lower than the single electron ionisation energies of inner valence (IV) and even some outer valence electrons. This is particularly important for carbon compounds, whose inner valence bands derive from the C2s atomic orbitals and have binding energies between 20 and 30 eV. If ionisation by single hole creation can occur at an energy above the double ionisation potential, it opens a new pathway to double ionisation akin to the usual inner shell Auger effect. This can greatly increase the intensity of double photoionisation.

We order the molecules in this chapter mainly in the usual way by number of atoms, then by molecular weight, but put closely related molecules together where possible.

Double Photoionisation Spectra of Molecules. John H. D. Eland and Raimund Feifel.
© John H. D. Eland and Raimund Feifel, 2018. Published in 2018 by Oxford University Press.
DOI 10.1093/oso/9780198788980.001.0001

5.1 Ethylene (ethene), C_2H_4

The ethylene molecule is planar with D_{2h} symmetry and has the electron configuration

$$1a_g{}^2 1b_{1u}{}^2 2a_g{}^2 2b_{1u}{}^2 1b_{2u}{}^2 3a_g{}^2 1b_{3g}{}^2 1b_{3u}{}^2 \ldots\ldots\ldots\ldots\ldots\ldots {}^1A_g.$$

The binding energies of the four outer orbitals from the photoelectron spectra [1, 2] are 10.5, 12.5, 14.5 and 15.8 eV. The $C_2H_4{}^{2+}$ dication is known to possess some bound or metastable states from the appearance of this ion in mass spectra, and has been studied extensively both in experiments and by high-level theory. In its ground state, from removal of two electrons from the outermost orbital, the dication is calculated [3] to be non-planar and to have an extended C—C bond. These differences between the neutral molecule and the dication mean that vertical double ionisation to the ground state will produce a broad and complex spectral band. The lowest excited states, which arise from removal of one electron from each of the outermost orbitals, are expected to be planar but to involve strong vibronic coupling to the ground state, further compromising any possibility of vibrational structure. In fact, only vibrationally unresolved bands are seen in the Auger spectrum [4], double charge transfer (DCT) spectra [5–8] and in the TOF-PEPECO spectrum [9] shown as Fig. 5.1.1.

To interpret the spectrum we turn to numerical calculations [7, 9, 10], particularly the most recent ones by Gaire et al. [10], who also carried out PEPECO measurements at 40.5 eV photon energy by the COLTRIMS technique (cold target recoil ion momentum spectroscopy [11]). Comparison of the spectrum with the calculated state energies strongly indicates that the first band, with its peak at about 31.5 eV, includes the ground

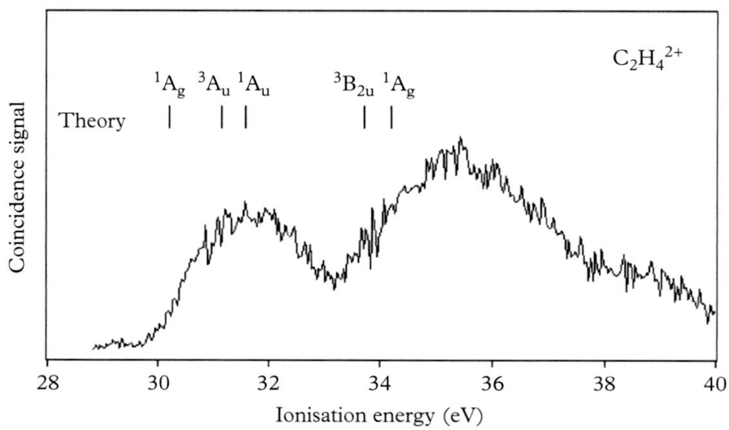

Figure 5.1.1 *TOF-PEPECO spectrum of ethylene at 40.81 eV. Vertical ionisation energies to the five lowest states calculated by Gaire et al. [10] are indicated. Six more states have been calculated in the 34–37 eV range.*

state and the first two excited states. Fournier et al. [8] argue, on the basis of the calculated equilibrium structures, that the ground state should appear as an extended band in a vertical ionisation process whereas the low-energy excited states individually should give narrower bands. This interpretation is corroborated by the DCT spectra, where the singlet and doublet states can be seen separately [7, 8], and it seems to provide a good interpretation of the first band in our TOF-PEPECO spectrum. The second band, which peaks near 35 eV, evidently also contains contributions from several states, as many as eight according to the calculations. This pile-up of states is consistent with the width and intensity of the band. A third band is hinted at in the spectrum, with peak possibly near 39 eV.

In contrast to our spectrum, Gaire et al. [10] obtained a two-electron coincidence spectrum with a very strong band at 30.2 eV and a very weak one at about 35 eV. The energy spacing between the bands is the same as in our spectrum (Fig. 5.1.1), but the peak energies are lower and the relative intensities are quite different. We are at a loss to explain these discrepancies at present.

References

[1] G.R. Branton et al., *J. Chem. Phys.*, **52** (1970) 802.
[2] D. Holland et al., *Chem. Phys.*, **219** (1997) 91.
[3] E. Ohrendorf et al., *J. Chem. Phys.*, **91** (1989) 1734.
[4] R.R. Rye et al., *J. Chem. Phys.*, **69** (1978) 1504.
[5] J. Appell et al., *J. Phys. B: At. Mol. Opt. Phys.*, 7 (1974) 406.
[6] C. Benoit and J.A. Horsley, *Mol. Phys.*, **30** (1975) 557.
[7] W.J. Griffiths and F.M. Harris, *J. Mass Spectrom. Ion Processes*, **122** (1992) 321.
[8] P.G. Fournier et al., *Chem. Phys.*, **178** (1993) 581.
[9] J.H.D. Eland, *Inst. Phys. Conf. Ser.*, **183** (2004) 115.
[10] B. Gaire et al., *Phys. Rev. A*, **89** (2014) 013403.
[11] J. Ullrich et al., *Rep. Prog. Phys.*, **66** (2003) 1463.

5.2 Butadiene (*trans*-buta-1,3-diene), C_4H_6

Butadiene is one of the simplest molecules to have two conjugated double bonds and has the structure CH_2=CH—CH=CH_2. As the *trans* isomer is centrosymmetric, its valence orbital structure is conventionally given as

$$5a_g{}^2 5b_u{}^2 6a_g{}^2 6b_u{}^2 7a_g{}^2 1a_u{}^2 1b_g{}^2 \ldots\ldots\ldots\ldots {}^1A_g.$$

In the photoelectron spectrum [1, 2] the first two bands at 9 and 11.5 eV are attributed to the two π-orbitals $1b_g$ and $1a_u$, while the third and fourth bands are at 12.2 and 13.4 eV. Double ionisation of this molecule had previously been investigated by a combination of Auger spectra, double charge transfer (DCT) spectra and Green's function theory [3]. The vertical double ionisation energies have also been calculated by a complete active space-SCF (CASSCF) method in connection with a study of the dissociation of the doubly charged ions [4] and by a multi-configuration quasi-degenerate perturbation theory approach in connection with ionisation by strong-field few-cycle infrared laser pulses [5].

The onset of the double ionisation spectrum in Fig. 5.2.1 is at 24.5 eV and there are several distinct peaks. In the Green's function theoretical results, the number of calculated states [3] is high even in the low-energy region and suggests considerable congestion. The more recent calculations [4, 5] find that the ground state, derived from $1b_g{}^{-2}$ ionisation, is well separated in energy from all excited states, so we can confidently assign the first band, with peak at 25.1 eV, as this ^1A_g state. The following band at 26.5 eV correlates with the first triplet state observed in the DCT spectrum [3], which

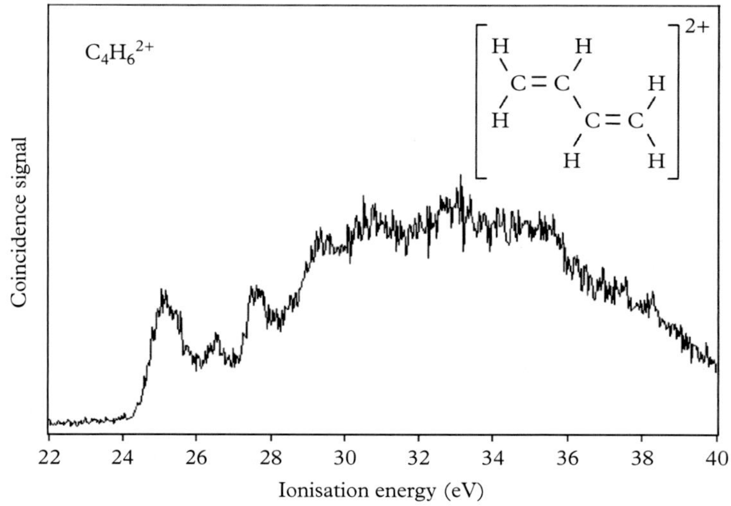

Figure 5.2.1 *TOF-PEPECO spectrum of butadiene taken at 40.81 eV photon energy.*

we identify with the triplet state from $1a_u^{-1}1b_g^{-1}$ ionisation at 1.79 eV excitation energy as calculated by Zhao et al. [5]. All the calculations suggest that the next band, which peaks at 27.6 eV, contains contributions from more than one state, most probably the triplet and singlet states from $7a_g^{-1}1b_g^{-1}$ ionisation. At higher energies, the calculated states become congested and most of them have multiple configurations. Positions of the singlet states measured by DCT [3], when shifted to make the first one coincide with our first band at 25.1 eV, agree reasonably with peaks in the TOF-PEPECO spectrum [6]. It is clear that a planar centrosymmetric structure is not retained for long by the dication in any state [4], and some structural changes may affect the vertical double ionisation spectrum.

Suggested identification of double ionisations of butadiene

Energy (eV)	Ionisation	Multiplicity
25.1	$1b_g^{-2}$	1
26.5	$1a_u^{-1}1b_g^{-1}$	3
27.6	Mixed	1 and 3
29.3	Mixed	3 and 1

References

[1] M. Beez et al., *Helv. Chim. Acta*, **56** (1973) 1028.
[2] J. Dannacher et al., *Chem. Phys.*, **51** (1980) 189.
[3] C.-M. Liegener et al., *Phys. Rev. B*, **46** (1992) 11295.
[4] S. Oghbaie et al., *J. Chem. Phys.*, **143** (2015) 114309.
[5] A. Zhao et al., *J. Phys. Chem. A*, **120** (2016) 3233.
[6] J.H.D. Eland, *Inst. Phys. Conference Ser.*, **183** (2004) 115.

5.3 Five-membered rings: pyrrole, furan and thiophene

The compounds pyrrole, furan and thiophene have planar five-membered ring structures containing four carbon atoms and one of NH, O or S. All have six π-electrons and are aromatic according to the Hückel $4n + 2$ rule (modern formulation [1]), and in their chemical properties. The valence orbitals of furan, each one doubly occupied, are

$$3b_2 4b_2 6a_1 7a_1 1b_1(\pi) 5b_2 6b_2 8a_1 9a_1 2b_1(\pi) 1a_2(\pi) \ldots\ldots\ldots\ldots\ldots\ldots\ldots {}^1A_1.$$

The same orbitals are occupied in pyrrole and thiophene with very much the same energy ordering, the only major difference being that the innermost π-orbital lies at different relative positions in the three molecules. The ionisation energies for removal of a single electron from the outermost π-orbitals are 8.21 eV in pyrrole, 8.90 eV in thiophene, and 8.87 eV in furan [2–4]. It is worth noting that the outermost orbital, $1a_2$, has no density at all on the heteroatom because of its symmetry. A possible doubly ionised state derived from dominant $1a_2^{-2}$ ionisation might be expected to come at about the same energy in the spectra of all three compounds, and could appear only in C1s Auger spectra, not in those based on heteroatom vacancies.

The double ionisation spectra of these molecules had hardly been examined by conventional methods such as Auger spectroscopy before the first TOF-PEPECO work on thiophene and some substituted thiophenes appeared [5]. The gas phase Auger spectra of furan have been presented by Rennie et al. [6], but without detailed analysis. The C1s Auger spectrum indeed contains a band at low ionisation energy which is absent from the O1s Auger spectrum, and this was attributed to $1a_2^{-2}$ ionisation, at about 25.5 eV [6]. The first band in the TOF-PEPECO spectrum of furan in Fig. 5.3.1 has its peak at 25.9 eV and may include a contribution from such $1a_2^{-2}$ ionisation. The Green's function calculations that accompanied the thiophene TOF-PEPECO spectrum [5], on the other hand, indicate that all the states of the doubly charged molecule have mixed configurations, and the lowest ${}^1A_1$ state has more $2b_1^{-2}$ than $1a_2^{-2}$ ionisation character. The same calculations find a triplet state, ${}^3B_2$ from $2b_1^{-1}1a_2^{-1}$ ionisation, to be the ground state of doubly ionised thiophene. The lowest energy ionisations from all three molecules are almost certainly based mainly on electron removals from the two outer π-orbitals, conventionally giving four states: ${}^3B_2$ and ${}^1B_2$ from $2b_1^{-1}1a_2^{-1}$ ionisation, and two ${}^1A_1$ states from $1a_2^{-2}$ and $2b_1^{-2}$ ionisations. Because the electron vacancies, and thus charges, are delocalised over a larger area in $2b_1^{-2}$ than in $1a_2^{-2}$ ionisation, we believe that the highest energy state of this group is mainly from $1a_2^{-2}$. Then the similarity in general form of the lowest bands in the three spectra in Fig. 5.3.1 suggests that the relatively small second bands at 26.4, 26.9 and 26.2 eV in the three compounds can be ascribed to this $1a_2^{-2}$ ionisation. The higher energy bands in the spectrum of thiophene have been explained by the Green's function calculations [5] but are assigned to groups of orbitals, rather than to individual configurations.

The yields of stable doubly charged parent ions of the aromatic molecules have been measured by Hartman et al. [7] as a function of photon energy. They find double ionisation onsets of 24.2 eV for pyrrole and 25.18 eV for furan, values which agree with the rising slopes of the first bands in our spectra.

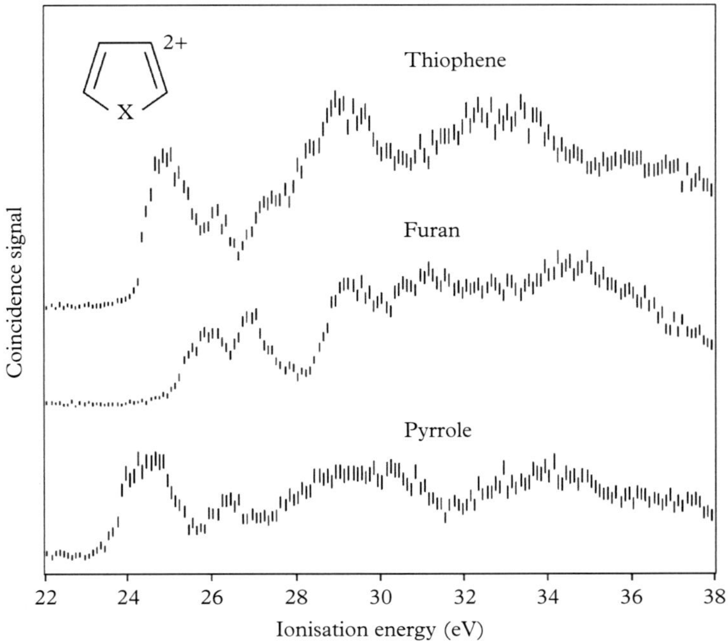

Figure 5.3.1 *TOF-PEPECO spectra of doubly ionised five-membered ring heterocycles at 40.81 eV photon energy. In pyrrole X= NH, in furan X = O and in thiophene X = S.*

Double ionisation energies (eV) of five-membered ring heterocycles

Molecule	Onset	Peaks of bands
Pyrrole	23.4	24.5, 26.4, 29, 34
Furan	25.0	25.9, 26.9, 29.2, 31, 34
Thiophene	24.25	24.9, 26.2, 27.4, 28.3, 29.1, 32

References

[1] J.D. Roberts, A. Streitweiser and C.M. Regan, *J. Am. Chem. Soc.*, **74** (1951) 4579.
[2] P.J. Derrick et al., *Int. J. Mass Spectrom. Ion Phys.*, **6** (1971) 191.
[3] W. von Niessen, W.P. Kraemer and L.S. Cederbaum, *J. Electron Spectrosc. Related Phenom.*, 8 (1976) 179.
[4] R.C.C. Perera and R.E. LaVilla, *J. Chem. Phys.*, **84** (1986) 4228.
[5] P. Linusson et al., *J. Chem. Phys.*, **129** (2008) 234303.
[6] E.E. Rennie et al., *J. Chem. Phys.*, **117** (2002) 6524.
[7] T. Hartman et al., *Phys. Rev. A*, **87** (2013) 063403.

5.4 Benzene, C_6H_6

Its hexagonal ring gives the benzene molecule D_{6h} symmetry and the valence electron configuration

$$2e_{2g}{}^4 3a_{1g}{}^2 2b_{1u}{}^2 1b_{2u}{}^2 3e_{1u}{}^4 1a_{2u}{}^2(\pi) 3e_{2g}{}^4 1e_{1g}{}^4(\pi) \ldots\ldots\ldots\ldots {}^1A_{1g}.$$

As the prototype aromatic molecule, benzene, and its ions have been studied extensively using all kinds of spectroscopic techniques. The photoelectron spectrum was first fully analysed by Jonsson and Lindholm [1] and it was extended later by Baltzer et al. [2] to include the inner valence region, with an important state (I^2A_{1g}) near 26 eV. The outer π-orbital is ionised at 9.25 eV, the first σ-orbital at 11.6 eV and the second π-orbital at 12.3 eV. A majority of both the experimental and theoretical work on benzene double ionisation has been about the structure, rearrangements and reactions, mainly dissociations, of the dications. Much of this work included measurements of the double ionisation energy, but approaches to the double ionisation spectrum began with the Auger spectrum [3] and its interpretation [4], double charge transfer (DCT) spectra [5, 6], the TPEsCO spectrum [7] and finally TOF-PEPECO [8].

The wide-range TOF-PEPECO spectra in Fig. 5.4.1 contain three distinct bands in the low-energy region and two or three less-distinct bands at higher energy. The spectrum at 90 eV closely resembles the Auger spectrum [3] in band intensities and positions

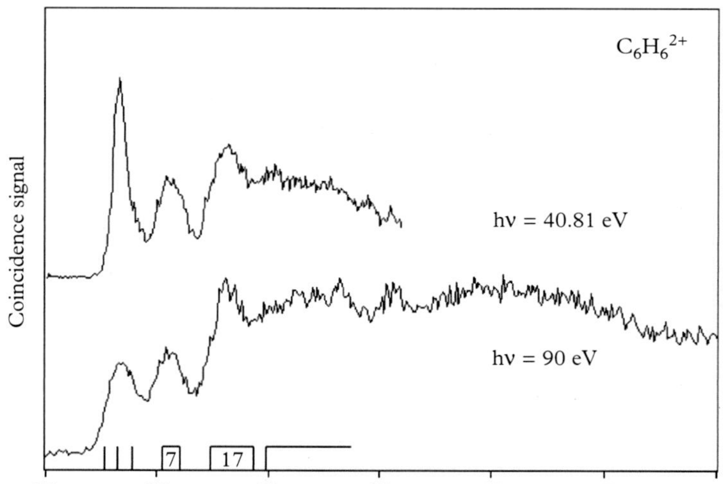

Figure 5.4.1 *TOF-PEPECO spectra of benzene at 90 and 40.81 eV photon energy. Spectra taken with other photon energies do not reveal any new bands. Theoretical state energies and groups of states from ref. [4] are indicated below the spectra as explained in the text.*

and so is well interpreted by the Green's function calculations of Tarantelli et al. [4], once the theoretical state energies are shifted by +1.3 eV to conform to experiment. Because there are so many (226) theoretically predicted states, bands in the spectra can be interpreted in terms of groups rather than individual electronic states. Although the positions of the bands are not affected by changes of photon energy, their relative intensities are affected and are changed even more if low-energy electrons are excluded in the data analysis. The reason is that the inner valence states near 26 eV can undergo Auger decay only to the low-energy final states of the dication, necessarily producing one high ($h\nu - 26$ eV) and one low ($26 - 25 = 1$ eV) energy electron. This sort of process, involving two fully allowed single-electron steps is more intense than direct double photoionisation, which requires electron correlation.

To examine the spectrum with better resolution, we used a number of lower photon energies to reduce the electron energy. The best results came from using the higher energy component of the NeII doublet at 26.91 eV (near 461 Å), producing the spectra shown in Fig. 5.4.2 for $C_6H_6^{2+}$ and $C_6D_6^{2+}$.

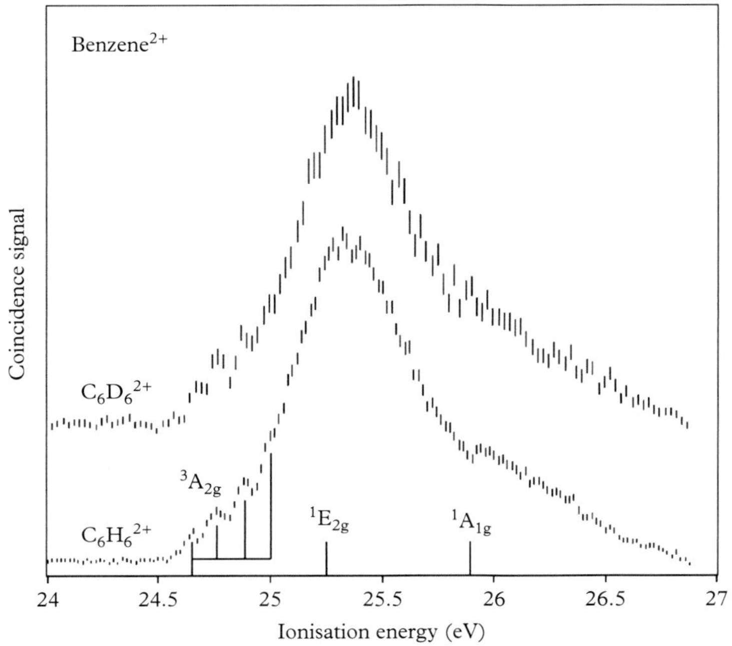

Figure 5.4.2 *TOF-PEPECO spectra of benzene and perdeuterobenzene at 26.91 eV photon energy. The theoretical energies [4] of the first three electronic states, adjusted to fit the first one to the threshold, are indicated as vertical bars, with the vibrational structure in the triplet ground state added. All the three electronic states in this first band come from $1e_{1g}^{-2}$ ionisation.*

At the photon energy used for the spectra in Fig. 5.4.2, the 26 eV inner valence state of singly charged benzene can be populated and indirect double ionisation by its Auger decay cannot be distinguished from direct double photoionisation, whatever the choice of electron energies. Indirect double photoionisation via this inner valence state may therefore contribute to the spectrum. In any case, the vibrational structure, though weak, is distinct both here and in the TPEsCO spectrum [7] and gives a clear adiabatic double ionisation threshold of 24.65 eV. The apparent vibration intervals of about 120 meV are consistent with the ring-breathing mode (ν_2) of benzene, as are the barely perceptible shifts to higher energy and smaller intervals on deuteration. The fact that these vibrations appear so clearly is consistent with the idea that the dication in its ground state retains the molecule's hexagonal structure, as has recently been demonstrated by infrared photodissociation of He-tagged complexes at low temperature [9]. Benzene dications in some of the higher states undergo rearrangements to more stable isomers, and ultimately dissociate by ring opening and hydrogen atom migration to form CH_3^+ + $C_5H_3^+$ and other charge-separated products [10].

Double ionisation energies (eV) of benzene

Molecule	State	Peaks of vibration levels
C_6H_6	$^3A_{2g}$	24.65, 24.76, 24.88, 24.99
C_6D_6	$^3A_{2g}$	24.68, 24.76, 24.87, (24.98)
		Peaks of bands
C_6H_6	$^1E_{2g}$	25.3
	$^1A_{1g}$	25.9
	Mixed	27.6, 30.1, 35.3, 37.5

References

[1] B.-Ö. Jonsson and E. Lindholm, *Ark. Fiz.*, **39** (1969) 65.
[2] P. Baltzer et al., *Chem. Phys.*, **224** (1997) 95.
[3] K. Siegbahn et al., in *ESCA Applied to Free Molecules*, North Holland, Amsterdam, 1969.
[4] F. Tarantelli et al., *J. Chem. Phys.*, **86** (1987) 2201.
[5] J. Appell et al., *J. Phys. B: At. Mol. Opt. Phys.*, 7 (1974) 406.
[6] W.J. Griffiths and F.M. Harris, *Chem. Phys.*, **157** (1991) 299.
[7] R.I. Hall et al., *Chem. Phys.*, **187** (1994) 125.
[8] J.H.D. Eland, *Chem. Phys.*, **345** (2008) 82.
[9] J. Jašik, D. Gerlich and J. Roithová, *J. Am. Chem. Soc.*, **136** (2014) 136.
[10] P.J. Richardson, J.H.D. Eland and P. Lablanquie, *Org. Mass Spectrom.*, **21** (1986) 289.

5.5 Hexafluorobenzene, C_6F_6 and toluene, C_6H_5—CH_3

These two benzene relatives illustrate the effects of substitution and reduction of symmetry on the double ionisation spectra. The hexafluorobenzene molecule has D_{6h} symmetry like benzene itself and the valence electron configuration is

$$1b_{1g}{}^2 6e_{1u}{}^4 1a_{2g}{}^2 6e_{2g}{}^4 2b_{2u}{}^2 2a_{2u}(\pi)^2 2e_{1g}(\pi)^4 \ldots\ldots\ldots {}^1A_{1g}.$$

In toluene, C_7H_8, neglecting the effect of the out-of-plane methyl hydrogen atoms, the symmetry is reduced to C_{2v}, and the outermost π-orbital and all deeper e-symmetry orbitals are converted into two non-degenerate orbitals. The five outermost orbitals are

$$1b_1(\pi)^2 1a_1(\sigma)^2 7b_2(\sigma)^2 1a_2(\pi)^2 2b_1(\pi)^2 \ldots\ldots\ldots {}^1A_1.$$

The first ionisation energy of toluene is 8.83 eV, about 0.4 eV lower than that of benzene, while that of hexafluorobenzene at 9.91 eV exceeds that of benzene by 0.66 eV. Because of the larger numbers of atoms and orbitals, both double ionisation spectra are congested and consist mainly of broad bands. The inner valence Auger effect *via* C2s-based orbital ionisations is significant for toluene, and for C_6F_6 it is strong *via* F2s-based states.

Toluene

The double ionisation spectrum of toluene had not been examined by Auger or double charge transfer (DCT) spectroscopy before the original TOF-PEPECO study [1], but its double ionisation energy had been measured by electron impact and other methods [2]. Its ionisation and subsequent decay have long been of interest because of the possibility that the product of H^+ loss from the dication, or H loss from the monocation, $C_7H_7{}^+$, might rearrange to the aromatic tropylium ion structure as a seven-membered ring [3]. According to calculations [2], the three lowest energy doubly ionised states should all arise from removal of two π-electrons, ${}^1A_1$ from $2b_1{}^{-2}$, ${}^3B_2$ and ${}^1B_2$ from $2b_1{}^{-1}1a_2{}^{-1}$. It seems likely that these three states are the main contributors to the first band in the spectrum of toluene dications in Fig. 5.5.1, whose onset is at 23.6 eV. The calculations also predict an extremely congested Frank–Condon envelope in population of the lowest state ${}^1A_1$, but less-congested envelopes with some distinct vibrational structure in population of the two states from $2b_1{}^{-1}1a_2{}^{-1}$ ionisation. The spectra in Fig. 5.5.1 do indeed show clear vibrational structure near the peak of the first composite band, particularly at the lower photon energy. The structure is not visible in equivalent spectra of fully deuterated toluene [1]. We therefore conclude that the theoretical analysis of Roithová et al. [2] is correct and interpret the spectra accordingly.

Hexafluorobenzene

The double ionisation spectrum of hexafluorobenzene was first investigated in a detailed study combining Auger spectroscopy, DCT spectroscopy and theory in 1993 [4]; this

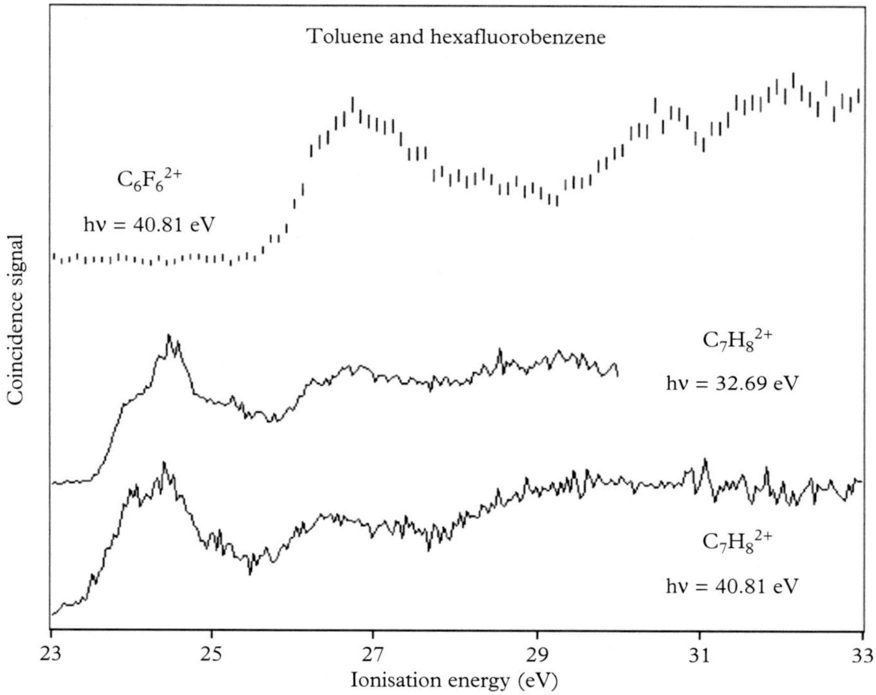

Figure 5.5.1 *TOF-PEPECO spectra of toluene at 40.81 and 32.69 eV photon energy (lines) and hexafluorobenzene at 40.81 eV (error bars). There are no distinct features at higher ionisation energies, just continuous bands of increasing relative intensity.*

work was later updated with more refined DCT measurements [5]. The TOF-PEPECO spectrum was later described but not presented in full [6]. Only broad bands are seen in the TOF-PEPECO spectrum, with an onset at 25.8 eV. The high resolution DCT results of Bayliss et al. [5] enable us to interpret the first band, with peak at 26.8 eV, as comprising the four calculated states $^3A_{1g}$, $^3E_{1u}$, $^1A_{1g}$ and $^1A_{2g}$ in that order. All these states arise from configurations where the electrons have been removed from the π-orbitals, $2e_{1g}$ and $2a_{2u}$.

It is interesting to compare the shifts in double and single ionisation on substitution in benzene.

The single ionisations shift by –0.4 and +0.66 eV relative to benzene itself while the double ionisations shift by −1.05 and +1.2 eV, essentially twice as much. Greater shifts in double as compared with single ionisation are a general phenomenon and reflect the importance of Coulomb repulsion energy, and hence of the distance to which the charges can separate.

Double ionisation energies (eV) of toluene from TOF-PEPECO

State	Onset	Peaks
$^1A_1$	23.6	25.93
$^{1\text{ or }3}B_2$		24.33, 24.48, 24.59 (vibration levels?)
Others		25.1, 26.7

Double ionisation energies (eV) of hexafluorobenzene from TOF-PEPECO

State/configuration*	Onset	Peaks
$^3A_{1g}$	25.8	
$2e_{1g}^2$		26.8
$^3E_{2u}$?		28.5
Others		30.5, 32.0

*Details of these and other states can be found in ref. [5].

References

[1] R.D. Molloy and J.H.D. Eland, *Chem. Phys. Lett.*, **421** (2006) 31.
[2] J. Roithová et al., *J. Chem. Phys.*, **122** (2005) 0943056.
[3] J. Roithová et al., *J. Phys. Chem. A*, **110** (2006) 2970.
[4] W.J. Griffiths et al., *J. Chem. Soc., Faraday Trans.*, **89** (1993) 1637.
[5] M.A. Bayliss et al., *Int. J. Mass Spectrom. Ion Processes*, **164** (1997) 241.
[6] J.H.D. Eland, *Chem. Phys.*, **294** (2003) 171.

5.6 Pyridine, C_5H_5N and the three diazines, $C_4H_4N_2$

The heterocyclic aromatic compounds containing nitrogen atoms in six-membered rings are similar to benzene in structure, but very different in chemical behaviour. Their molecular orbital structure, demonstrated in their photoelectron spectra [1–3], is also significantly different as the outermost orbitals are in all cases nitrogen lone pairs, rather than π-orbitals. In the case of pyridine, the exact ordering is in doubt [2] but only because the first π-orbital and the nitrogen lone-pair orbital are so close in energy that their photoelectron bands overlap. The outer orbitals, from which all the low-lying doubly ionised states arise, are either π-orbitals or combinations of nitrogen lone pairs. The lone-pair orbitals lie in the plane of the rings and have some C and C—H character beside the dominant N-contribution, but are more localised than the π-orbitals. Because of the Coulomb repulsion energy in dications, localisation versus delocalisation has a major effect on the double ionisation spectra.

The double ionisation of pyridine was first analysed by Rye et al. [4], with particular emphasis on the N-Auger spectrum. The C-Auger spectrum was reported to have a first peak at a binding energy of about 26 eV, not present in the N-Auger spectrum, which was attributed to a double vacancy in the outermost π-orbital, $1a_2$, which has no density at the N atom. This attribution accords well with the first band in the TOF-PEPECO spectrum in Fig. 5.6.1, with peak at 25.7 eV and onset at about 25.1 eV. The second band is more intense and extends from 26.2 to 27.8 eV, and is certainly composite. Rye et al. present no calculated energies for other final states involving a hole in $1a_2$, but predict the existence at low energy of states from four configurations based on the $7a_1$ (lone-pair), $2b_1(\pi)$ and $1b_1(\pi)$ orbitals. The lowest energy double ionisation as calculated is $7a_1^{-1}2b_1^{-1}$, which is probably a component of the second band or possibly, as a triplet state, of the first band. Other contributions which we expect to exist there will come from $7a_1^{-1}1a_2^{-1}$ and $1a_2^{-1}2b_1^{-1}$. More complete theoretical calculations are needed to interpret this spectrum, which has not been published before.

Of the diazines, it is only for pyrimidine that high-level calculations have been carried out on double ionisation. The Auger spectra were measured and interpreted by Storchi et al. [5], partly because of interest in pyrimidine as a building block for the nucleobases and ultimately for DNA. Calculation by the Green's function method gave very good fits to both the N- and C-Auger spectra, but as the calculations were for Auger spectra, only singlet final states were reported. All the final states were reported to have mixed two-hole configurations, but the lowest singlet state was entirely based on the nitrogen lone-pair orbitals, which are $5b_2$ and $7a_1$ in C_{2v} symmetry. The simple argument based on the Coulomb repulsion energy indicates that the lowest energy dication states must have one charge on each nitrogen atom. The binding energy for this lowest singlet state, as deduced from the Auger spectrum, depends on an assumed value for the energy of the parent state with a single hole in the N1s orbital. Storchi et al. [5] used a theoretical value, but if we take the experimental value reported by Vall-llosera et al. [6], the binding energy comes to 27 eV, in good agreement with the position of the first band peak in our TOF-PEPECO spectrum of pyrimidine at 26.8 eV. The band has an unclear onset at about 26.1 eV and seems to be composite, with a component producing a shoulder at

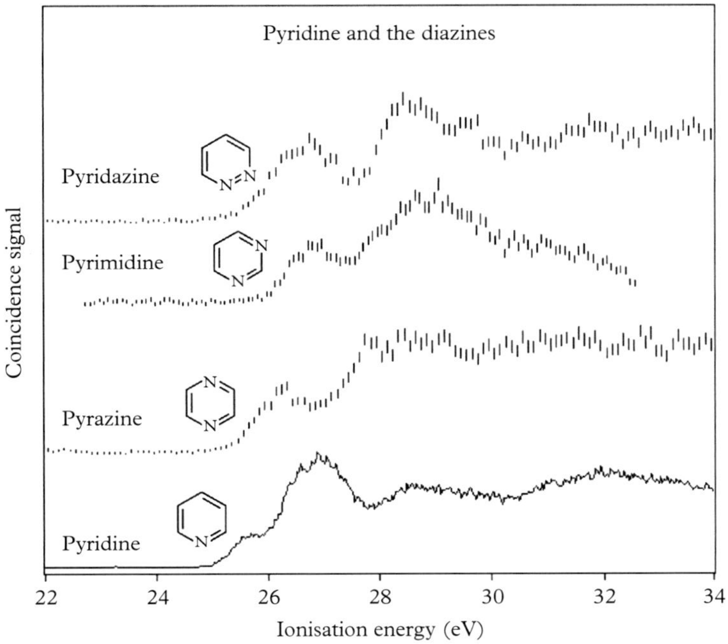

Figure 5.6.1 *TOF-PEPECO spectra of pyridine (full line) and the diazines (error bars) taken at 40.81 eV photon energy except for pyrimidine, which was taken at 32.69 eV.*

26.5 eV; if real, this might be a triplet state. The higher singlet states calculated by Storchi et al. [5] to be important in the N-Auger spectrum all have mixed configurations, with binding energies around 29.5, 32 and 33.2 eV in three groups corresponding to the Auger lines. We suggest that the second main band in the TOF-PEPECO spectrum, with peak at 28.7 eV, probably derives from triplet and singlet states with the same configurations involving only lone-pair and π-orbitals. From these four orbitals alone 16 final two-hole dication states can be generated, 10 singlets and 6 triplets. If the first band in the spectrum comes from pure lone pairs as suggested, the second can comprise a further 12 states.

For the remaining diazines, pyrazine and pyridazine, there are no published calculations of the dication states known to us, so we are thrown back onto qualitative arguments. Substitution of nitrogen atoms into the benzene ring has two opposing effects. Firstly, the ionisation energies are raised by the higher nuclear charge of N atoms (electron-withdrawing effect). This explains why pyridine has a lower first double ionisation energy than any of the diazines. Secondly, the nitrogen lone pairs are less tightly bound than the π-orbitals, and ionisation from them produces the lowest energy states. Amongst the diazines we expect the lowest double ionisation energy to be found where the N atoms, and hence the charges, can be furthest apart. This explains why pyrazine has the lowest double ionisation onset, at 25.5 eV, and the lowest energy at the peak of the first band at 26.2 eV.

Since double photoionisation generally populates both triplet and singlet final states, we suggest that the first band in each spectrum contains at least the triplet and singlet from removal of one electron from each N atom. The second band contains the multiple states from ionisation out of lone pairs and the outer π-orbitals, including the removal of both lone-pair electrons from a single atom. The third and higher bands, which merge into a continuum, contain states from ionisation out of the inner π-orbitals and σ-orbitals.

Like all compounds containing carbon, pyridine and the diazines have inner valence single ionisation bands at binding energies near 24 eV coming from states of mainly C2s atomic character, but in contrast to most aromatic molecules these bands lie just below, rather than above, the molecular double ionisation thresholds. This means that the inner valence Auger effect has only a minor impact on their spectra.

Double ionisation energies (eV) of the azabenzenes from TOF-PEPECO

Molecule	Onset	Peaks (shoulders)
Pyridine	25.1	25.6, 26.9, 32
Pyrazine	25.5	(25.9), 26.2, 27.7
Pyrimidine	26.1	(26.5), 26.8, (28.0), 28.7
Pyridazine	25.6	(25.8), 26.6, 28.4, 32

References

[1] R. Gleiter, E. Heilbronner and V. Hornung, *Helv. Chim. Acta*, **55** (1972) 28.
[2] W. von Niessen, G.H.F. Diercksen and L.S. Cederbaum, *Chem. Phys.*, **10** (1975) 345.
[3] A.W. Potts et al., *J. Phys. B: At. Mol. Opt. Phys.*, **36** (2003) 3129.
[4] R.R. Rye et al., *J. Chem. Phys.*, **88** (1987) 4375.
[5] L. Storchi et al., *J. Chem. Phys.*, **129** (2008) 154309.
[6] G. Vall-llosera et al., *J. Chem. Phys.*, **128** (2008) 044316.

5.7 Naphthalene and azulene, $C_{10}H_8$

Naphthalene is the prototype of large polynuclear aromatic molecules which occur widely in many environments, largely because of their great stability as neutral molecules and as both singly and doubly charged ions. With two *cata*-condensed benzene rings it is the first of the acenes, while its isomer azulene with one five-membered and one seven-membered ring is a less stable compound, most unusual for its bright blue colour. In such large molecules as these, the electronic structure inevitably needs many close-spaced molecular orbitals for its description, but conjugated π-orbitals are always outermost. For naphthalene, the π-orbitals were identified in the HeI photoelectron spectrum very early [1, 2] and a full theoretical description of the electronic structure by Green's function calculations was given later [3]. Critically for its double photoionisation behaviour, inner valence states based on $C2s^{-1}$ ionisation are expected around 24 eV; the theoretical work of Deleuze et al. [3] predicts the existence of many such states at binding energies up to 28 eV. For azulene, the HeI photoelectron spectrum is also known [2, 4] but there are no high-level theoretical calculations available. The ionisation energies most relevant to interpretation of the double ionisation spectra are the first two π-orbital binding energies. These are 8.15 eV ($1a_u^{-1}$) and 8.88 eV ($2b_{1u}^{-1}$) for naphthalene, and 7.43 eV ($2a_2^{-1}$) and 8.50 eV ($3b_1^{-1}$) for azulene. All the π-orbitals are non-degenerate.

In the TOF-PEPECO spectra of both azulene and naphthalene in Fig. 5.7.1, there are prominent bands at low ionisation energy, followed by a rapidly rising signal at higher

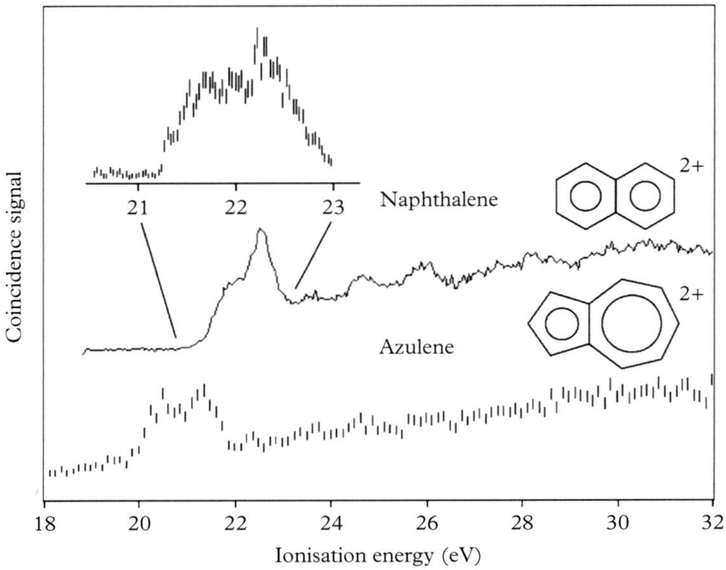

Figure 5.7.1 *TOF-PEPECO spectra of azulene and naphthalene at 40.81 eV photon energy and of naphthalene at 23.08 eV (HeIβ) on an expanded scale.*

ionisation energies. These spectra were extracted from the raw data with inclusion of all electron pairs, so including a large contribution from indirect double ionisation via the inner valence states based on C2s ionisation. If this process is excluded by accepting only electrons of near-equal energies, the relative intensities of the low-energy bands decrease very markedly. According to our own low-resolution HeII photoelectron spectra, measured concurrently with the TOF-PEPECO spectra [5], the inner valence bands are intense for naphthalene, with a peak around 22.2 eV and extending up to 28 eV as predicted [3]. For azulene, the inner valence states appear to form no distinct band, but tail off gently to about 24.5 eV; in both cases there is significant inner valence intensity above the double ionisation thresholds. The same situation holds for all larger aromatic compounds, and means that indirect ionisation by the inner valence Auger effect is a major mechanism in double ionisation by all impact techniques.

To interpret the spectra in the absence of high-level theory, we turn to the simple orbital model with Coulomb energy. The low-energy bands in the spectra of the two compounds both have two main components, with separations between their peaks (ca 0.6 and 0.9 eV) which almost match the splittings in binding energy between the respective outer π-orbitals (0.73 and 1.1 eV). We therefore assign the lowest energy portions of the bands as the singlet states from removal of both electrons from the outermost orbitals, and the second portions of the bands to the triplet and singlet states where one electron is removed from each of the first two π-orbitals. The first double ionisation band of naphthalene has a more complex structure which may be partly vibrational when examined at high resolution with low photon energy.

Double ionisation energies (eV) for naphthalene and azulene

Molecule	Ionisation	Onset	First peak	Later peaks or shoulders
Naphthalene	$1a_u^{-2}$	21.28	21.52	21.7, 22.2, 23.5, 24.6, 25.9
Azulene	$2a_2^{-2}$	20.2	20.7	21.3, 22.4

References

[1] P.A. Clark, F. Brogli and E. Heilbronner, *Helv. Chim. Acta*, **55** (1972) 1415.
[2] R. Boschi, E. Clar and W. Schmidt, *J. Chem. Phys.*, **60** (1974) 4406.
[3] M.S. Deleuze, A.B. Trofimov and L.S. Cederbaum, *J. Chem. Phys.*, **115** (2001) 5859.
[4] J.H.D. Eland and C.J. Danby, *Z. Naturforsch.*, **23a** (1968) 355.
[5] J.H.D. Eland, *Chem. Phys.*, **345** (2008) 82.

5.8 Quinoline, C_9H_7N and biphenyl, $C_{12}H_{10}$

The spectra of these two molecules are presented with minimal comment as exemplars of the rather bland spectra to be expected from larger aromatic molecules and to illustrate how the inner valence Auger effect can dominate double ionisation spectra.

Figure 5.8.1 shows the low-resolution photoelectron spectrum of quinoline at 40.81 eV measured at the same time as the TOF-PEPECO spectrum. The normal spectrum is extracted from the raw electron data by accepting all electron pairs, and shows the total population of final states produced by photoionisation. When the electron–electron coincidence map is examined it is found to be extremely hollow, meaning that it is dominated by pairs of very unequal energy, one very low, here less than about 4 eV, and one much higher. These pairs are formed by photoelectrons of 13–17 eV creating inner valence states between 23 and 27 eV, as seen in the photoelectron spectrum, and the Auger (or autoionisation) electrons of 0–4 eV forming dications. To obtain a spectrum free of this inner valence Auger effect we select only electron pairs with energy differences less than 5 eV at the double ionisation threshold, and proportionally smaller energy

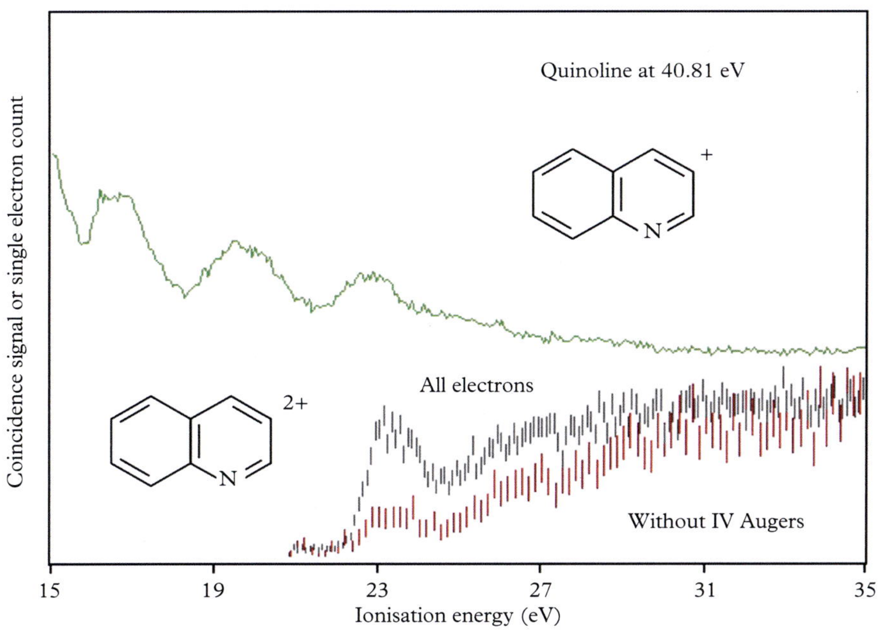

Figure 5.8.1 *Spectra of singly (green line) and doubly charged quinoline measured concurrently at 40.81 eV photon energy. To eliminate the inner valence (IV) Auger effect, electron pairs with large energy differences were excluded in the analysis (red error bars) and the resulting data were normalised to match the complete TOF-PEPECO spectrum (black error bars) in the high energy region.*

differences at lower total pair energy. This rather elaborate selection is needed to avoid severe distortion of the spectrum which would result if a simple minimum electron energy limit was set.

It is clear from comparison of the two TOF-PEPECO spectra that the energy states of the doubly charged ions are formed in high abundance by the inner valence Auger effect, but are completely dominated by dication states of higher energy in direct double photoionisation. In polycyclic aromatic molecules like quinoline, the molecular framework is so rigid that only minor changes in geometry are expected between the neutral molecule, the inner valence ionised monocation, and the final dication, so the shapes of the low-energy bands (Franck–Condon envelopes) are probably very similar in direct and indirect photoionisation.

In double photoionisation of biphenyl, Fig. 5.8.2, very much the same phenomena due to an inner valence Auger effect occur even more markedly. The main inner valence ionisation band has peaks at 22.2 eV, well above the onset of double ionisation, which lies in the region of 20.5 eV. There is uncertainty about the true onset in this case, because unlike quinoline, biphenyl is not a rigid molecule. In its neutral form the rings are not coplanar, but they probably become so in its singly and doubly ionised incarnations. For this reason the Franck–Condon envelopes for direct and indirect double photoionisation

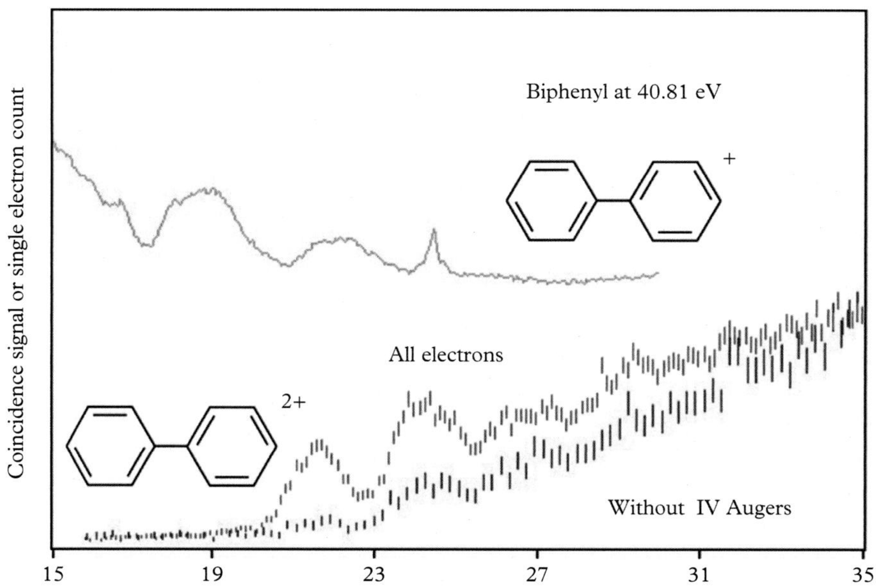

Figure 5.8.2 *Photoelectron and TOF-PEPECO spectra of biphenyl with the same conventions as in Fig. 5.8.1. The sharp peak in the photoelectron spectrum at 24.5 eV is from ionisation of He leaking from the lamp.*

are likely to be significantly different. In particular, if the intermediate singly charged inner valence state has the same geometry as the doubly charged final state, the Auger transition is likely to reach the lowest part of the final dication potential energy surface, while this lowest level may be inaccessible to a direct vertical transition from the neutral ground state. The different profiles of the lowest energy bands in the direct and indirect TOF-PEPECO spectra of biphenyl in Fig. 5.8.2 suggest that this is indeed the case, though the direct spectrum in Fig. 5.8.2 has too few counts for complete certainty. Effects of this sort must always be expected where there are inner valence single ionisation bands above the double ionisation onsets, particularly in floppy molecules and molecules which dissociate easily.

No prior work on the double ionisation of these two molecules is known to us, and these TOF-PEPECO spectra have not been published elsewhere.

Double ionisation energies (eV) of quinoline and biphenyl

Molecule	Onset	Peaks (shoulders)
Quinoline	22.4	23.2, (26)
Biphenyl	20.5	21.6, 24.1, 26.3

5.9 Molecules with very low ionisation energies

As molecules get bigger, their ionisation energies, both single and double, generally become lower. But as they get bigger they also become progressively less volatile and less convenient as targets for experiments on free molecules in the gas phase. There are a few easily available small molecules with exceptionally low ionisation energies and reasonable volatility, and these are of interest to experimentalists as they allow high-order processes to be examined in the laboratory using simple target gas production and relatively low-energy light sources. The double ionisation spectra of four such molecules are presented in Fig. 5.9.1, at least three of them with double ionisation thresholds below the energy of the widely used HeIα line (21.218 eV).

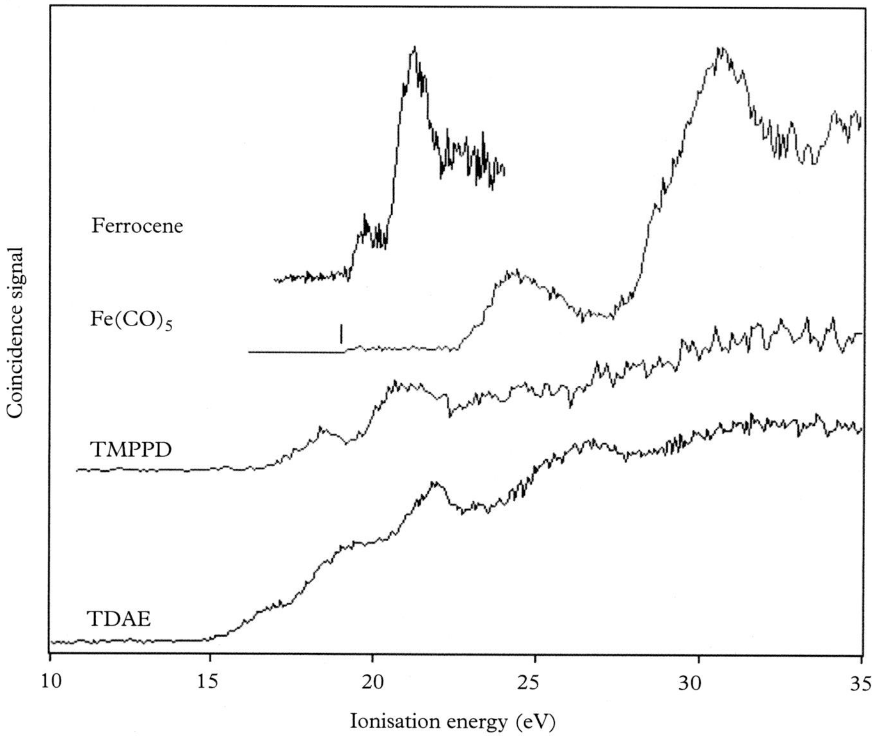

Figure 5.9.1 *TOF-PEPECO spectra showing very low double ionisation energies: TDAE (tetrakis(dimethylamino)ethylene) at 48.37 eV photon energy; TMPPD (N,N,N′,N′-tetramethyl* p-*phenylene diamine) at 40.81 eV, iron pentacarbonyl at 40.81 eV, and ferrocene (bis-*η^5*-cyclopentadienyl iron,* $Fe(C_5H_5)_2$*) at 26.91 eV. The weak onset marked by a vertical line near 19.3 eV for* $Fe(CO)_5$ *may be caused by an impurity.*

The lowest molecular ionisation energy known to us is that of tetrakis(dimethylamino)ethylene (TDAE), whose adiabatic single ionisation energy is reported to be as low as 5.4 eV [1, 2]. This is a fascinating compound, a foul-smelling oily liquid, which emits a bright light from complex chemical reactions immediately on contact with oxygen or air. We have recorded its double ionisation spectrum at several photon energies, but there is no major change in shape of the spectrum on going from 23.08 to 51 eV photon energy. In both single and double ionisation, changes in molecular geometry are expected between neutral and ionised forms, producing broad bands in the spectra. The onset of double ionisation is at about 15.0 eV, undoubtedly a record low value for a molecule, but whether this is ionisation to an intact molecular ion must be open to some question. The double ionisation energy is so low that there are many singly ionised states of both inner valence and even pure outer valence character, which can act as intermediates in Auger processes. In some such intermediate states, substantial changes in geometry or actual dissociation are likely before the second electron is ejected. A slightly more pleasant compound to handle is N,N,N′,N′-tetramethyl *p*-phenylene diamine (TMPPD or TMPD) whose single ionisation threshold is about 1 eV higher [1]. Its double ionisation onset is at 16.9 eV, and in this case, while Auger processes are almost certainly present, dissociation before double ionisation is perhaps less likely. The zero-valent iron compounds, $Fe(CO)_5$ and ferrocene would have almost identical measured double ionisation energies at 19.25 and 19.35 eV if the very low intensity onset of the $Fe(CO)_5$ spectrum at about 19.3 eV were a true molecular double ionisation of the compound. It is striking that the apparent double ionisation thresholds of the two molecules would then be almost the same, while the single ionisation thresholds are quite different, 7.98 eV for $Fe(CO)_5$ [3] and 6.85 eV for ferrocene [4]. The apparent double ionisation from both molecules at about 19.3 eV is certainly not ionisation of bare iron atoms from their ground state, which would require 24.1 eV. Using the simple model of twice the first ionisation energy plus a Coulomb energy (Appendix 3), we estimate the double ionisation threshold of $Fe(CO)_5$ as between 21.5 and 24 eV, depending on the location of the charges. Furthermore, the lowest appearance energy of a doubly charged product in a study of dissociative double photoionisation of $Fe(CO)_5$ was estimated as 25 eV [5], so both estimates point to the first strong band in the TOF-PEPECO spectrum, which we tentatively conclude is real double ionisation of the $Fe(CO)_5$ molecule. If this is true, the double ionisation threshold observed at 19.3 eV must be due to an impurity in the sample, perhaps a higher carbonyl such as $Fe_3(CO)_{12}$. On the other hand, all the different forms of density-functional theory calculation of the double ionisation energy of $Fe(CO)_5$ by Chong [6] make it about 20.9 eV, and if the true vertical DIE is 24.2 eV, the deviation between experiment and theory is greater in this case than in any other. So perhaps this case should remain open for the present.

Low double ionisation energies (eV)

Molecule	Onset	Peaks (shoulders)
TDAE	15.5	(16.6), (19.4), 22.0, 26.6
TMPPD	16.9	18.5, 20.7
Ferrocene	19.35	19.7, 21.2
$Fe(CO)_5$	22.9	24.2, (28.9), 30.75

References

[1] Y. Nakato et al., *Chem. Phys. Lett.*, **9** (1971) 615.
[2] B.H. Smith and R.N. Compton, *J. Phys. Chem. A*, **118** (2014) 7288.
[3] G. Distefano, *J. Res. Nat., Bur. Standards*, **74A** (1970) 233.
[4] R. Gleiter, M.C. Bohm and R.D. Ernst, *J. Electron Spectrosc. Related Phenom.*, **33** (1984) 269.
[5] S. Hsieh and J.H.D. Eland, *Int. J. Mass Spectrom. Ion Processes*, **167/168** (1997) 415.
[6] D.P. Chong, *J. Chem. Phys.*, **128** (2008) 084112.

Chapter 6

Mainly aliphatic molecules

6

Mainly aliphatic molecules

Several aliphatic molecules containing a single carbon atom and a single substituent group have appeared in earlier chapters; here we show double photoionisation spectra of a few larger, mainly aliphatic molecules, including parts of two homologous series. The lowest double ionisation energies are higher in aliphatic and saturated molecules as compared with their unsaturated and aromatic counterparts, so here the inner valence Auger effects are less prevalent, at least in smaller molecules. As molecules get larger the double ionisation energies become lower and the spectra, like regular photoelectron spectra, become less and less structured. The loss of structure is more marked in double ionisation than in single, because there are many more states and the vast majority of dication states in large molecules are dissociative. As little detailed analysis of the individual spectra is possible, we present the molecules of this chapter in groups rather than individually.

In this chapter, molecules are ordered more thematically than strictly by size. We start with four homologous iodides and three homologous alcohols. Then we take some individual molecules with different substituent groups and proceed to a few larger molecules.

Double Photoionisation Spectra of Molecules. John H. D. Eland and Raimund Feifel.
© John H. D. Eland and Raimund Feifel, 2018. Published in 2018 by Oxford University Press.
DOI 10.1093/oso/9780198788980.001.0001

6.1 Monoiodides, $C_nH_{2n+1}I$

In the double ionisation spectrum of methyl iodide (Fig. 4.7.2), repeated here in Fig. 6.1.1, it is clear that in the lowest energy state both charges reside on the iodine atom. Even in ethyl iodide this is evidently no longer true; the energetic imperative to keep the charges as far apart as possible overcomes the relatively low binding energy of the iodine lone pairs. There are no sharp lines even where they might be expected at slightly higher energy to show an excited state with a pure iodine-only double hole configuration. The same pattern is repeated in the larger molecules, where the lowest double ionisation onset decreases steadily as size increases. There is a relatively large decrease in the lowest double ionisation energy on going from n-propyl iodide (1-iodopropane) to n-hexyl iodide (1-iodohexane), accompanied by a large diminution in the intensity near the onset. These two features rather strongly suggest that the lowest double ionisation energy seen in the electron pair spectrum is not a vertical process at the geometry of

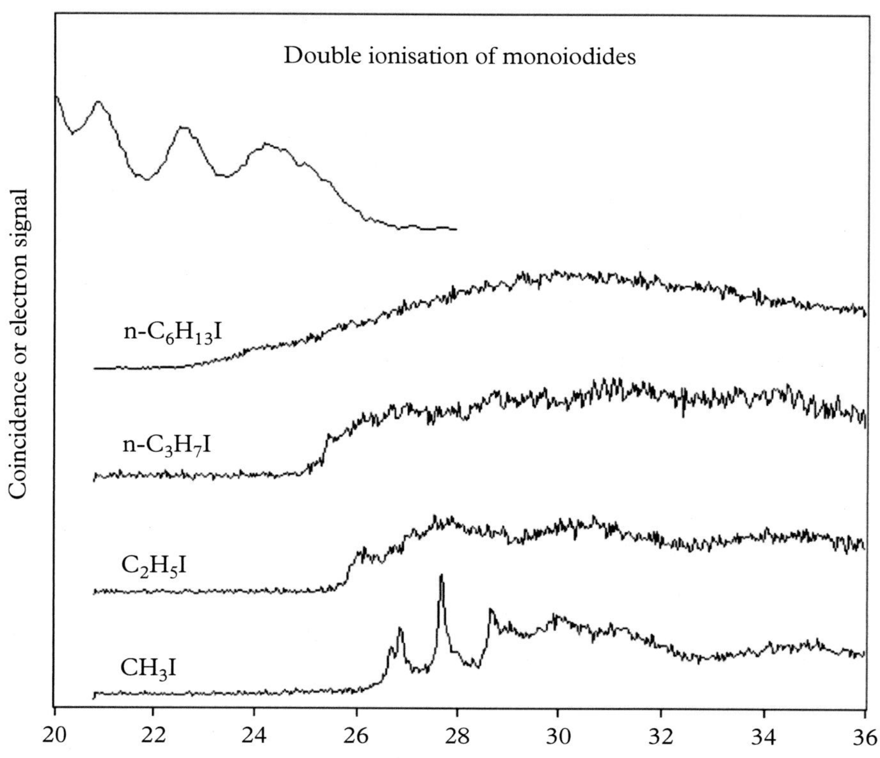

Figure 6.1.1 *Double photoionisation spectra of the iodides indicated, all taken by TOF-PEPECO at 40.81 eV photon energy, with part of the single ionisation spectrum of n-hexyl iodide measured at the same photon energy (uppermost curve).*

the neutral molecule. The double ionisation process near threshold is probably mainly indirect, as the single ionisation spectrum (uppermost curve in Fig. 6.1.1) shows strong inner valence bands covering the same energy range. There is a strong likelihood of dissociation, or at least of substantial rearrangement in the intermediate singly ionised inner valence states before Auger decay.

The TOF-PEPECO spectra of the iodides were analysed by Molloy et al. [1] in connection with an empirical rule for molecular double ionisation energies (Appendix 3).

Approximate onsets of double ionisation for the iodides (not vertical DIEs)

CH_3I	26.5 eV
C_2H_5I	25.8 eV
n-C_3H_7I	25.2 eV
n-$C_6H_{13}I$	22.9 eV

Reference

[1] R.D. Molloy et al., *Chem. Phys.*, **335** (2007) 49.

6.2 Aliphatic alcohols, $C_nH_{2n+1}OH$

The double ionisation of the alcohols from methanol to n-propanol, shown in Fig. 6.2.1, exhibits the same characteristics as the spectra of the homologous iodides.

As the alcohol molecules get bigger, the onset of double ionisation moves to lower energy and the spectra become more and more bland. The double ionisation spectra

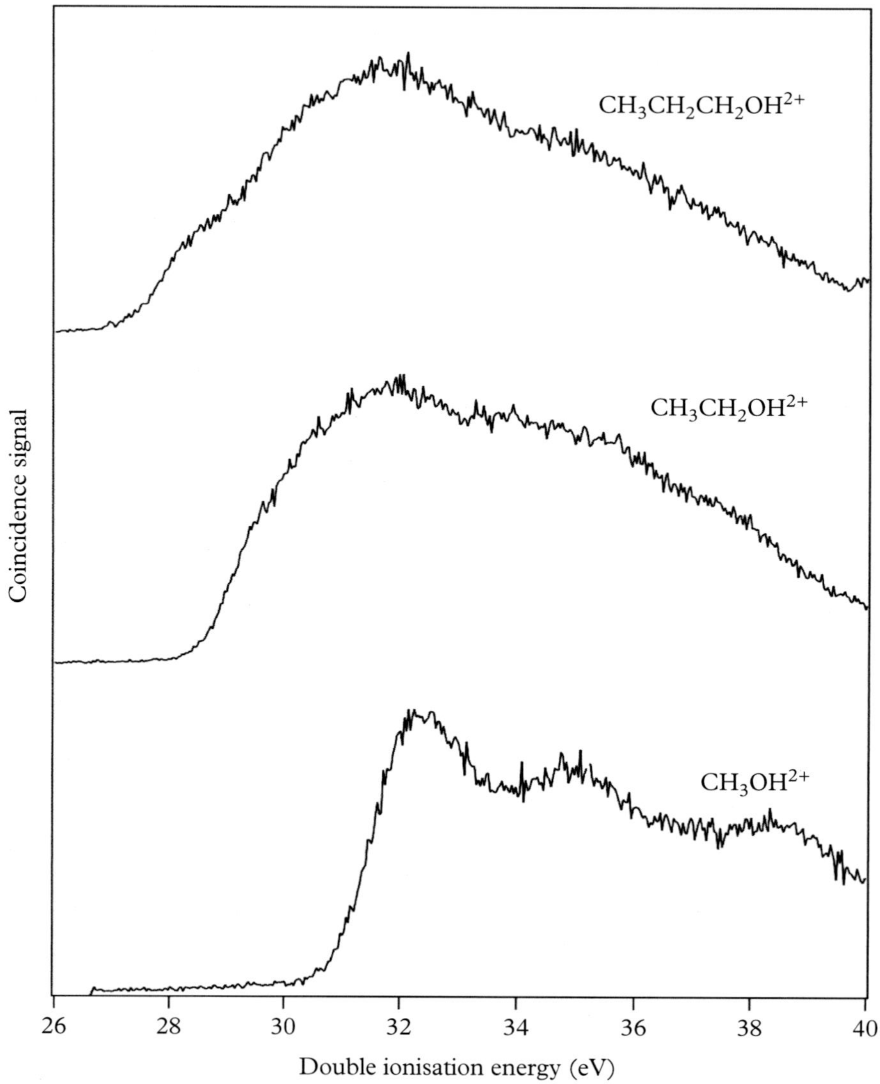

Figure 6.2.1 *TOF-PEPECO spectra of three aliphatic alcohols, methanol, ethanol and n-propanol, at 40.81 eV photon energy.*

of these molecules were measured by Linusson et al. [1] and compared with vertical double ionisation energies calculated using high-level theory (CASSCF). As the spectra have no very marked structure, the help of theory was vital in the attempt to identify features corresponding to the lowest vertical double ionisation. The vertical double ionisation energies so obtained are given in the list below, together with the double ionisation onsets, which may not be related to vertical transitions. The empirical model mentioned in connection with the iodides (see Appendix 3) was found to give model intercharge distances in reasonable agreement with the detailed calculations.

Onsets and first vertical double ionisation energies of alcohols

	Onset (eV)	Vertical DIE (eV)
Methanol	30.5	32.1
Ethanol	28.2	29.6
n-Propanol	27.0	28.2

Reference

[1] P. Linusson et al., *Phys. Rev. A*, **80** (2009) 032516.

6.3 Acetonitrile, CH_3CN and methyl mercaptan, CH_3SH

The double ionisation spectra of acetonitrile (methyl cyanide) and methyl mercaptan (methanethiol) are shown in Fig. 6.3.1 for comparison with their unmethylated congeners HCN and H_2S.

Comparison between the simple hydrides and the methylated compounds in Fig. 6.3.1 illustrates both the loss of resolvable vibrational structure and the very large decrease in double ionisation energy that follows from methylation. The decreases of more than 4 eV in both these two cases are quite incompatible with simple inductive effects, which are largely responsible for the much smaller decreases in the first single ionisation energies (1.0 eV for the sulphides and 1.7 eV in the cyanides). The major factor, in addition to inductive lowering of the S3p electron binding energy, must be the extension of the molecular framework, allowing the charges to separate to a greater distance, reducing the Coulomb repulsion energy in the doubly charged molecules. The

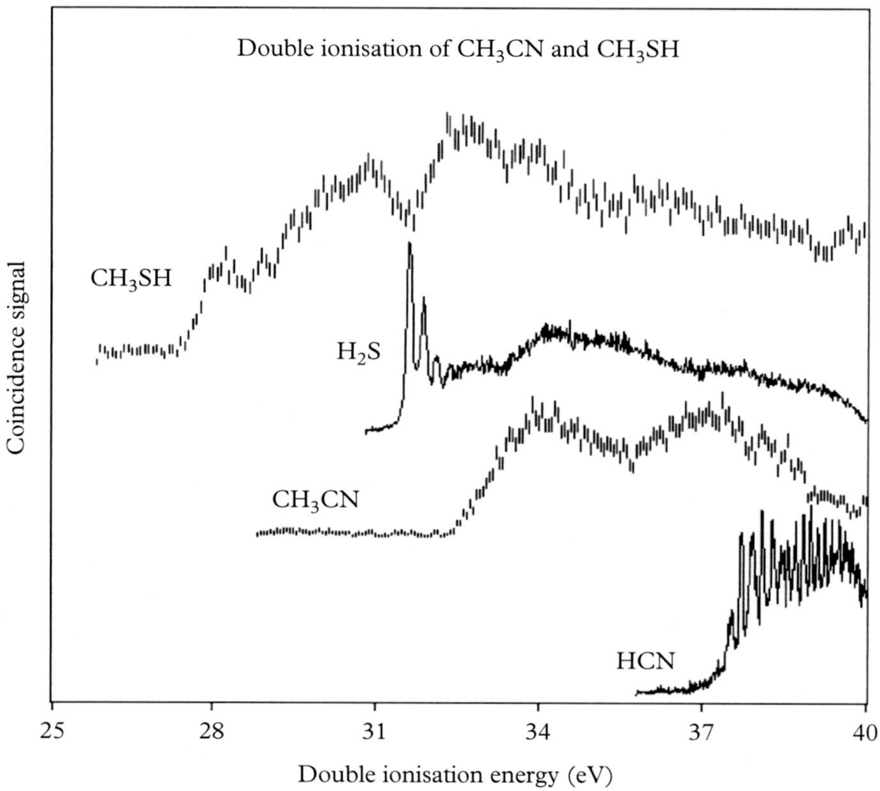

Figure 6.3.1 *TOF-PEPECO spectra of CH_3CN and CH_3SH compared with those of HCN and H_2S, all measured at 40.81 eV photon energy.*

loss of vibrational structure is mainly due to spectral congestion, where changes in geometric structure upon double ionisation may also play a part. Both the carbon and nitrogen Auger spectra of acetonitrile were reported and analysed by Rye and Houston [1] in terms of an orbital model. The first band in both Auger spectra, which is relatively stronger in the nitrogen Auger spectrum, is attributed to the removal of two π-electrons from the CN group. This band corresponds roughly in energy to the first band in the present TOF-PEPECO spectrum.

Onsets and double ionisation energies at peaks in the spectra

	Onset (eV)	**DIEs at peaks (eV)**
CH_3CN	32.6	34.0, 37.0
CH_3SH	27.5	28.2, 28.9, 30.9, 32.4

Reference

[1] R.R. Rye and J.E. Houston, *J. Chem. Phys.*, 75 (1981) 2085.

6.4 Acetaldehyde, CH_3CHO and acetone, CH_3COCH_3

Acetaldehyde (ethanal) and acetone (2-propanone) were some of the first polyatomic molecules whose double ionisation was studied in detail by Auger spectroscopy and related theoretical calculations [1].

The double photoionisation of these two related molecules produces rich spectra with a number of distinct peaks, but the theoretical analysis [1] shows that all the peaks arise from multiple final states of the doubly charged ions. The lowest energy bands in both spectra seem to arise mainly from ionisation local to the carbonyl groups, as expected since the carbonyl π-orbitals are outermost. This is a rather typical situation for large molecules, where high-level theory can reproduce the spectra, but because of their complexity gives only limited insight into the nature of the states produced. The behaviour of the nascent doubly charged ions is then difficult to predict theoretically,

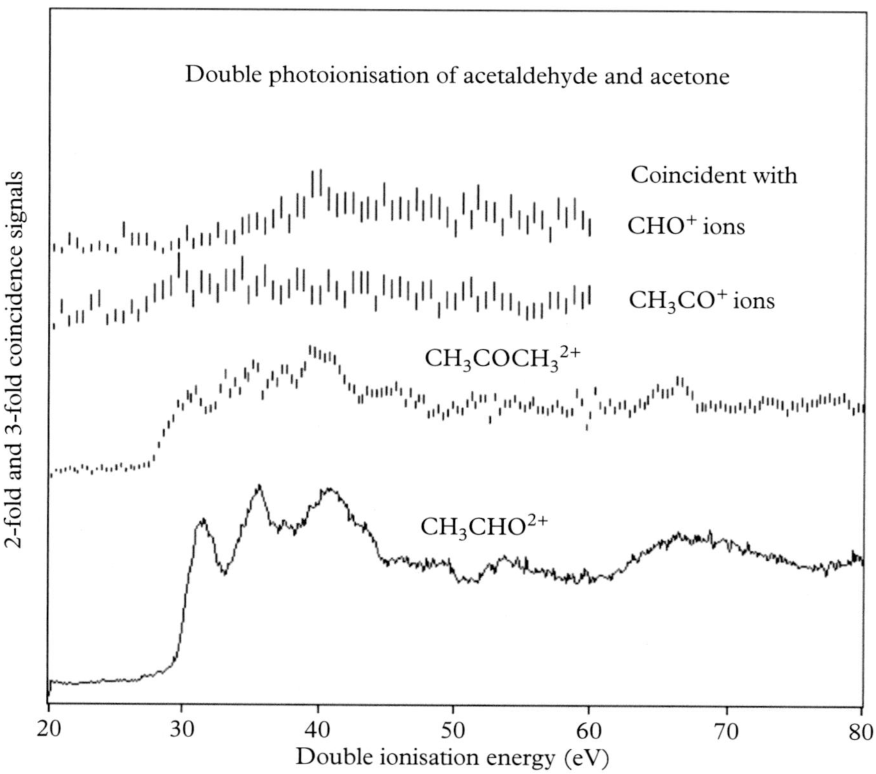

Figure 6.4.1 *TOF-PEPECO spectra of acetaldehyde at 95 eV and acetone at 90 eV photon energy, with selected double ionisation spectra of acetone in coincidence with CH_3CHO^+ and CHO^+ ions measured simultaneously and drawn with error bars above the overall double ionisation spectrum of acetone itself.*

but can be examined experimentally by additional techniques. The measurements on acetone shown in Fig. 6.4.1 were made using one such technique, by which the ions produced from the events giving each electron pair were also detected in coincidence. In this instance both electrons and ions were measured using a single magnetic bottle spectrometer, by applying a pulsed electric field to accelerate ions to the detector after the electrons had already left the source region [2]. The simplest and lowest energy accessible dissociation produces the ion pair CH_3^+ + CH_3CO^+; the acetyl ion is found accordingly in coincidence with the lowest energy states of the molecular dication. Formation of CHO^+ (with a variety of partners) is an abundant process despite the necessary rearrangement, but sets in only at higher energy. Where the electronic spectra are congested and there are many internal vibration modes, as in this case, it is reasonable to assume that after double ionisation (but before all dissociations except perhaps H^+ loss or H migration), energy flows freely between all the electronic states. There will be many crossings of the multidimensional potential energy surfaces with conical intersections where this can happen.

Onsets and double ionisation energies at peaks in the spectra

	Onset (eV)	**DIEs at peaks (eV)**
CH_3CHO	29.7	31.3, 35.6, 40.8, 53.8, 67.3
CH_3COCH_3	28.2	30.7, 35.4, 40.0, 66.4

References

[1] N. Correia et al., *J. Chem. Phys.*, **95** (1991) 5187.
[2] J.H.D. Eland et al., *Chem. Phys. Lett.*, **548** (2012) 90.

6.5 Three larger molecules

In this section we show the double photoionisation spectra of three larger molecules, TMMD (NN′-tetramethyl methylene diamine, $C_5H_{12}N_2$), COT (*cyclo*-octatetraene, C_8H_8), and norbornane (bicyclo[2,2,1] heptane, C_7H_{12}), none of which have yet been analysed in detail.

The compounds in Fig. 6.5.1 have very different chemical characters; TMMD is a bifunctional tertiary amine, while COT is an unsaturated hydrocarbon with four double bonds but little conjugation and no aromaticity because of its non-planar geometric structure and norbornane is a tricyclic saturated hydrocarbon. Only the amine TMMD shows significant structure in its spectrum, presumably because of the easily ionised lone pair electrons on the two well-separated N atoms. The first band in the TMMD spectrum with peak at 21.8 eV can be confidently assigned as a state in which the two charges reside on the two separate $-N(CH_3)_2$ groups, but not necessarily at the neutral ground-state geometry [1]. The prototype ammonia molecule becomes planar at equilibrium when singly ionised, so considerable geometry change in the same general

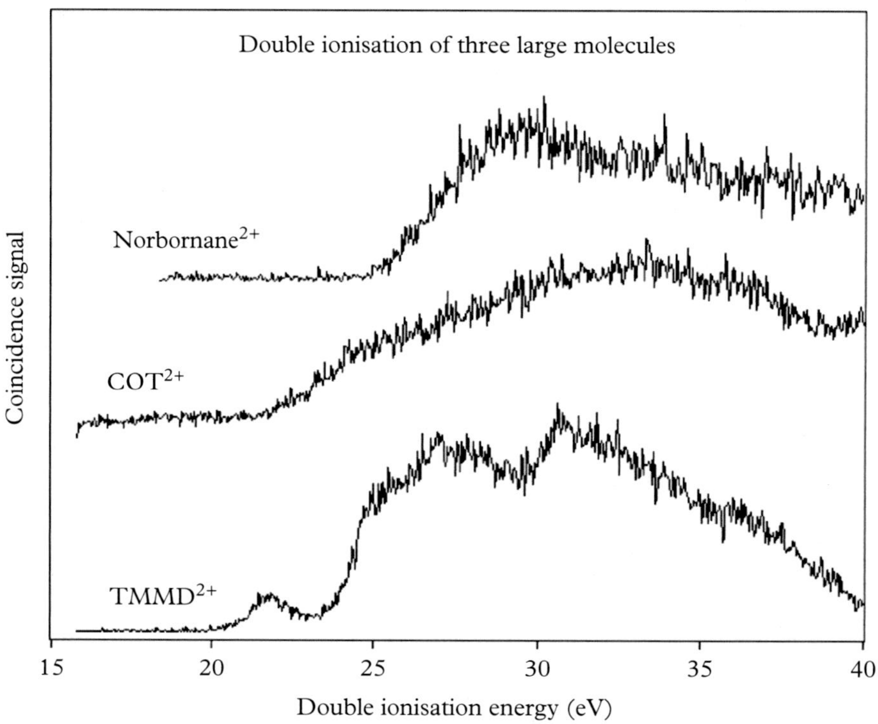

Figure 6.5.1 *TOF-PEPECO double photoionisation spectra of the three molecules indicated: TMMD and COT measured at 40.81 eV photon energy and norbornane measured at 48.37 eV.*

direction must be expected around each N atom in $TMMD^{2+}$. In addition, it is almost inevitable in a compound with such a low double ionisation threshold that a significant part of the double photoionisation is indirect, *via* inner valence singly ionised states based on C2s or N2s electron removal. The same applies to the double ionisation of *cyclo*-octatetraene [2], and with slightly less force, to the spectrum of norbornane [3]. Overall, with these spectra we emphasise that double ionisation energies of large molecules, whether measured by TOF-PEPECO, Auger spectroscopy or any other existing method, cannot be equated directly with vertical transitions defining a molecular ionisation energy, calculable reliably at the neutral molecule geometry.

Onsets and double ionisation energies at peaks in the spectra

	Onset (eV)	**DIEs at peaks (eV)**
TMMD	20.5	21.8, 25.3, 27.0, 36.6
COT	22.1	
Norbornane	25.2	29.5

References

[1] R.D. Molloy et al., *Chem. Phys.*, **335** (2007) 49.
[2] J.H.D. Eland, *Inst. Phys. Conference Ser.,* **183** (2004) 115.
[3] S. Knippenberg et al., *J. Phys. Chem.* A, **109** (2005) 4267.

Chapter 7

Core–core and core–valence double ionisation

7

Core–core and core–valence double ionisation

All the double ionisations so far in this book have involved removal of two electrons from the valence shells of molecules. In two higher-energy forms of double ionisation, one or both electrons can be removed from an inner shell instead, and the TOF-PEPECO technique is ideally, almost uniquely, suited to the study of such processes. Core–core double ionisation consists in removing both of two electrons from inner shells such as the 1s shells of light elements, which are not directly involved in bonding. For this reason, it is an almost purely atomic process and core–core ionisation spectra might ideally show a single line at a characteristic energy for ionisation from each distinct atom or pair of atoms. In reality, the sudden creation of inner shell charges will inevitably lead to changes in molecular structure, so a single band is produced, but may be hard to resolve into individual vibration levels by current techniques. The main interest focuses on the exact energies of the ionisation bands from the relatively few distinct core–core ionised states that can be produced in a small molecule. In core–valence double ionisation, by contrast, rich molecular spectra are produced because one electron comes from the occupied orbitals of the valence shell, so for a single core ionisation there are at least as many core–valence ionised states as there are different valence orbitals. Where the core electron can come from different atoms in the same molecule, different spectra are produced for each core charge localisation. The states produced in core–valence double ionisation, particularly of bare atoms, are also called *core shake-off satellites*, as their existence can be detected without spectral characterisation, in single-electron spectra, measured without coincidences. The states formed in both core–core and core–valence double ionisation have very short lifetimes as they rapidly undergo Auger decay, filling the core holes and ejecting further electrons. It is the capability to detect these extra electrons in coincidence with the two photoelectrons from formation of core hole states that makes TOF-PEPECO the most powerful method to identify and measure them.

In this chapter we first discuss core–core double ionisation briefly and then present the core–valence double ionisation spectra of a few interesting molecules.

In the context of core ionisation it is customary to use the very economical 'KLM' notation, which is universally used in Auger spectroscopy, as well as the chemists' familiar

Double Photoionisation Spectra of Molecules. John H. D. Eland and Raimund Feifel.
© John H. D. Eland and Raimund Feifel, 2018. Published in 2018 by Oxford University Press.
DOI 10.1093/oso/9780198788980.001.0001

1s, 2s, 2p and so on. We use the KLM notation sparingly in this chapter where its brevity is an advantage, so we include a short list of equivalents here.

K: 1s (principal quantum number $n = 1$)

L: orbitals with $n = 2$ in general

L_1: 2s, L_2: $2p_{1/2}$, L_3: $2p_{3/2}$

M: orbitals with $n = 3$, e.g. $M_{4,5}$: $3d_{3/2}$, $3d_{5/2}$

V: valence orbitals in general

7.1 Core–core double ionisation

Double vacancies in the 1s orbitals of atoms create 'hollow' atomic species, which were first detected in nuclear K-capture and as hypersatellite lines in X-ray emission spectra [1]. The formation of such species by double photoionisation attracted theoretical interest as an important testing ground for molecular structure models incorporating electron correlation [2]. Where molecules have a single central atom, the 'hollow' nomenclature can be extended to them, and indeed some of the first molecules with double 1s vacancies detected by TOF-PEPECO, CH_4 and NH_3 [3] were of this type. At almost the same time, the same technique was used to characterise double 1s vacancies in N_2, O_2, CO and CO_2 molecules, where there is only one atom of a given element or two identical atoms in symmetrically identical positions [4]. But in molecules like these which contain more than one atom with a 1s shell, a new possibility arises that two single 1s holes might be located on different atoms. This possibility was discussed in an early and prophetic paper by Cederbaum et al. [5] which demonstrated theoretically that double core–hole energies, particularly for states with core holes on different atoms, are a very sensitive probe of the molecular environment. This paper sparked further theoretical interest [6, 7] and was a strong motivation for subsequent TOF-PEPECO and other work.

The major difficulty in experimental study of double core holes by photoionisation is the very small magnitude of the cross-section for their formation. The cross-section for single-site double core hole (ss-DCH) production is generally about ten thousand times smaller than the cross-section for producing a single core hole. The probability of forming a two-site double core hole (ts-DCH) is likely to be a further thousand times less. The experimental technique has to be both very sensitive and very selective to distinguish DCHs from other processes and from background interference. This is achieved in TOF-PEPECO by a combination of the uniquely high electron collection efficiency with the selectivity conferred by multiple coincidence detection. When a DCH has been formed, the most intense subsequent decay is by two separate Auger processes, each emitting one further electron and so leading to a final quadruply charged final state. The Auger electrons have characteristic energies, which can be recognised in the electron spectra. The experimenter can then elect to filter the full data set by requiring the presence of all four electrons, or by taking just one Auger electron in addition to the two photoelectrons. The use of tuneable and intense synchrotron radiation from a storage ring to allow choice of a suitable photon energy is also a vital requirement. Even so, the resulting spectra are far from perfect, as Fig. 7.1.1 illustrates.

7.1.1 Single-site double core holes in 1s shells

The spectra in Fig. 7.1.1 have rather poor statistics and the peaks are rather wide. The statistics could be improved by longer runtimes, but the widths are only partly due to limited instrumental resolution. In the core doubly ionised states of both molecules, the X—H bond distance will certainly be different from the distances in the neutral ground states, so the unresolved peaks actually conceal bands of vibrational structure similar to the $1s^{-1}$ single ionisation bands. There must also be an intrinsic width from the inherent

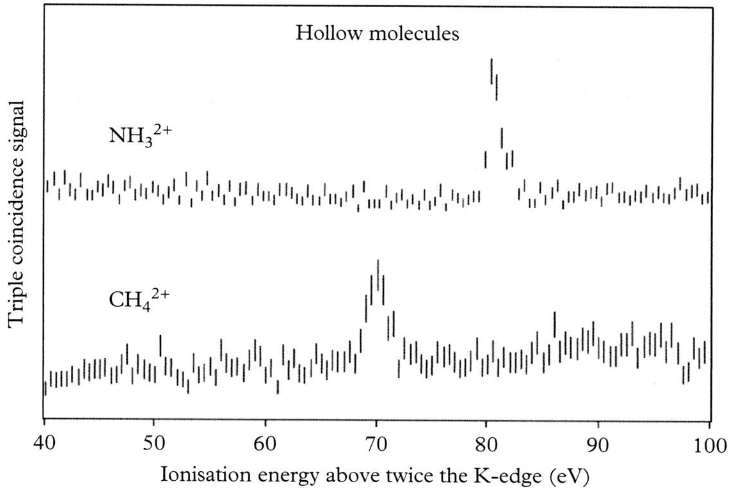

Figure 7.1.1 *Core–core double ionisation spectra of CH_4 and NH_3 from triple coincidences at 750 and 950 eV photon energy, respectively. Both spectra are plotted on a scale of the excess ionisation energy above twice the $1s^{-1}$ single ionisation energy.*

lifetime broadening, which is thought to be somewhat greater in K^{-2} decay than in K^{-1} decay of the same species [8]. This width will be of the order of a few hundred meV for both CH_4 and NH_3, and may make vibrational structure unmeasurable whatever the apparatus resolution.

The energies of K^{-2} states have by now been measured by coincidence and other techniques in a number of molecules containing the first-row elements C to F and at least one second row element (S) [3, 4, 9–11]. Elaborate *ab initio* calculations can be carried out to predict these ionisation energies with impressive accuracy [6] but for experimental design purposes and conceptual simplicity, a much cruder model can suffice. The 1s orbitals in molecules whose atoms also have occupied $n = 2$ (L) shells are essentially atomic. If they are modelled as hydrogenic, the most probable orbital radius r^* is inversely proportional to the nuclear charge, Z, $r^* = a_0/Z$ for a single electron where a_0 is the Bohr radius. If the excess ionisation energy above twice the energy for removal of a single electron from the 1s orbital is attributed to Coulomb repulsion between the notional positive holes, proportional to $1/r$, we can expect the energy for $1s^{-2}$ ionisation to be

$$\mathrm{IE}\left(1s^{-2}\right) = 2\,\mathrm{IE}\left(1s^{-1}\right) + kZ$$

where k is a constant and IE means ionisation energy. The constant k must be found experimentally, to somehow include the effects of reorganisation energy and correlation. The experimentally known $1s^{-2}$ ionisation energies fit remarkably well to this simple relationship with k= 11.7 eV, as illustrated in Fig. 7.1.2.

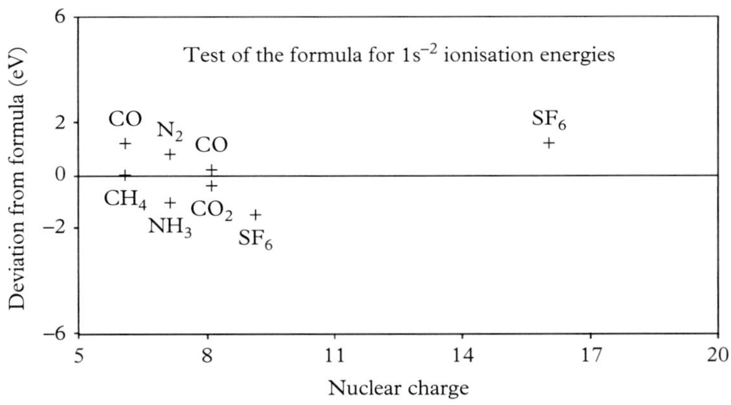

Figure 7.1.2 *Test of the formula* $IE(1s^{-2}) = 2\ IE(1s^{-1}) + 11.7Z$ *for predicting the energies of single-site double core holes in the 1s (K) shell. The experimental points (crosses) have uncertainties between* ± 0.5 *and* ± 2 *eV, comparable to the observed deviations.*

Two non-coincidence methods used to obtain some individual points in Fig. 7.1.2 are the production of the DCHs by two-photon ionisation using intense X-ray light from a free-electron laser [10], and examination of pre-edge structure by single electron spectroscopy [11, 12]. When two electrons are removed from the 1s shell, one of them may be excited to a normally unoccupied orbital rather than ejected completely. The states formed in this way can be located by single electron spectroscopy with a suitable X-ray light source, and their binding energies relative to the double core hole can be obtained rather precisely from high-level theory. This is therefore an indirect but effective experimental method to determine the energy of double core hole states.

7.1.2 Single-site double core holes in 2p and other shells

In molecules containing atoms from the second and later rows of the periodic table, we can in principle create double core ionisation by removing two electrons from any inner shell, or one electron from each of two different inner shells. The only such ionisations to have been examined so far are single-site double ionisations from 2p shells in molecules of second row atoms. These can be studied without coincidence techniques, because if a single hole is created in the 1s shell, its subsequent Auger decay proceeds at least partly by using one 2p electron to fill the 1s hole while a second 2p electron is ejected. The energy of the Auger electron emitted in this KLL Auger process, when combined with prior knowledge of the energy of the 1s hole state, gives the energy of the $2p^{-2}$ ionisation [13]. A limitation of this technique is that of the three electronic states possible from the $2p^4$ atomic configuration (^3P, ^1D and ^1S), only the ^1D state is generally recorded. The TOF-PEPECO technique is made highly selective by using fourfold or threefold coincidences, including one or both Auger electrons emitted in decay from the double core hole ($2p^4$) state, as well as the two electrons emitted in reaching the $2p^4$ configuration. The spectra gathered in this way show all three final states and provide precise binding energies [14].

Although we treat the S2p ionised states as atomic, they are affected by, and interact with the molecular environment, as is shown by shifts in their energies and the resolution of some $2p^{-1}$ ($^2P_{3/2}$) atomic states from single core ionisation into molecular components [15]. The main difference between the single and double core holes is that the chemical shifts are about twice as large in double core hole formation compared with single, confirming the prediction that these double core holes can provide a sensitive test of molecular environmental effects [5]. The inset in Fig. 7.1.3 demonstrates this by comparing the shifts in $S2p^{-2}$ against $S2p^{-1}$ levels, both relative to H_2S as standard, for a range of molecules from dimethyl sulphide to sulphur hexafluoride. Deviations from a simple straight line show that individual molecular characteristics play a part, and these can all be accurately modelled by molecular electronic structure theory [5–7, 14, 16].

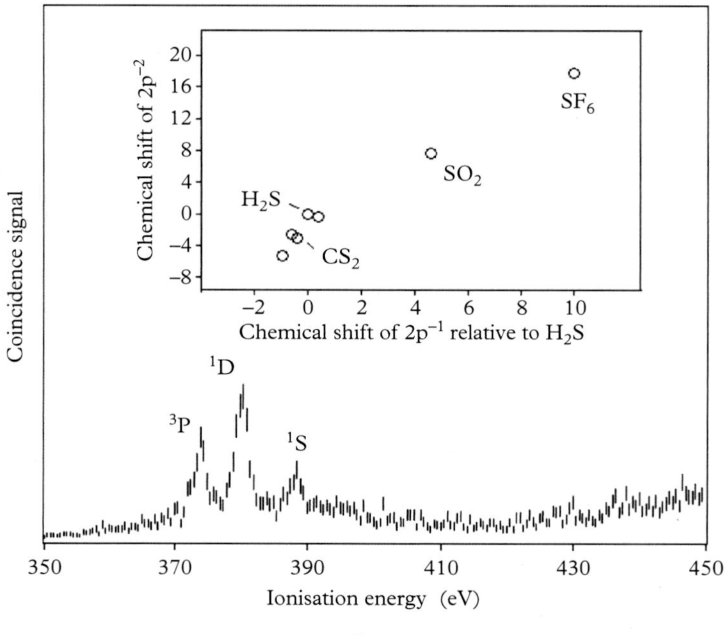

Figure 7.1.3 *Spectrum of the $S2p^{-2}$ states in H_2S taken at 500 eV photon energy using triple coincidences with an inset panel showing the relative chemical shifts of $2p^{-1}$ and $2p^{-2}$ levels for a range of molecules. The three points not identified in the inset panel, going from left to right, are for dimethyl sulphide, thiophene and carbonyl sulphide.*

7.1.3 Two-site double core holes

Because they are expected to be much more sensitive to chemical structure than their single-site cousins, great experimental efforts have been expended to detect doubly ionised states in which electrons have been removed from the core orbitals of two separate atoms. In single-photon impact, such states are formed mainly by the knock-out mechanism where the primary photoelectron from one core ejection causes a second core ionisation by an inelastic collision. Simple geometry dictates that this becomes less and less probable as distance between the atoms increases, but even for adjacent atoms the two-site process is about 100 times less likely than single-site double ejection. In principle, a two-photon process should be much more favourable, because in separate photon impacts the probability of core ionisation should be the same for all equivalent atoms in a molecule, independent of how many ionisations have already happened. Such experiments can be done only with the intense and very brief (few fs) X-ray pulses from a free-electron laser, because if a two-site DCH is to be formed, a second photon must do its work before the hole state created by a first photon has time to decay by an Auger process. A number of experimental campaigns by this route have claimed experimental detection of two-site DCHs [17–19], but the impact of the results is gravely diluted by the unavoidable presence of many parallel multiple ionisation processes producing a very congested background to the experimental spectra. The peaks due to ts-DCHs are recognised in practice only with considerable help from theory. The one absolutely unambiguous experimental measurement of two-site double core holes was accomplished using the TOF-PEPECO technique by Nakano et al. [12] on C_2H_2, C_2H_4, C_2H_6, CO and N_2. The results demonstrate the expected high sensitivity of the two-site DCH energy to bond length and confirm validity of the knock-out mechanism.

References

[1] J.P. Briant et al., *Phys. Rev. Lett.*, **27** (1971) 777.
[2] J. Hoszowska et al., *Phys. Rev. Lett.*, **102** (2009) 073006.
[3] J.H.D. Eland et al., *Phys. Rev. Lett.*, **105** (2010) 213005.
[4] P. Lablanquie et al., *Phys. Rev. Lett.*, **106** (2011) 063003.
[5] L.S. Cederbaum et al., *J. Chem. Phys.*, **85** (1986) 6513.
[6] M. Tashiro et al., *J. Chem. Phys.*, **132** (2010) 184302.
[7] O. Takahashi et al., *J. Phys. Chem. A*, **115** (2011) 12070.
[8] M. Chen, *Phys. Rev. A*, **44** (1991) 239.
[9] M. Nakano et al., *Phys. Rev. Lett.*, **110** (2013) 163001.
[10] N. Berrah et al., *Proc. Natl. Acad. Sci. U.S.A.*, **108** (2011) 16912.
[11] R. Feifel et al., *Scientific Reports*, 7 (2017) 13317.
[12] M. Nakano et al., *Phys. Rev. Lett.*, **111** (2013) 123001.
[13] R.N. Sodhi and R.G. Cavell, *J. Electron Spectrosc. Related Phenom.*, 32 (1983) 283.
[14] P. Linusson et al., *Phys. Rev. A*, **83** (2011) 022506.
[15] M. Poygin et al., *Phys. Rev. A*, **74** (2006) 012711.
[16] L. Hedin et al., *Chem. Phys.*, **439** (2014) 111.
[17] N. Berrah et al., *Proc. Natl. Acad. Sci.*, **108** (2011) 16912.
[18] P. Salén et al., *Phys. Rev. Lett.*, **108** (2012) 153003.
[19] M. Mucke et al., *New J. Phys.*, **17** (2015) 073002.

7.2 Core–valence double ionisation; generalities

While core–core double ionisation gives simple atomic lines and valence double ionisation produces highly congested spectra, core–valence (CV) double ionisation is a happy medium, giving spectra only a little more congested than the familiar one-electron photoelectron spectra. Although the cross-sections are small, the spectra fall in a relatively free region of electron energy space and so are not unduly difficult to measure.

We can formally imagine the process of core–valence double ionisation as happening in two ways. Valence photoionisation might take place first and the high energy photoelectron so produced could create a core hole by an inelastic collision with the core electrons. Alternatively, the core ionisation might happen first and then the low-energy ejected electron could produce a dication state by impact on the valence shell. Both pathways should produce spectra not too different in shape from the normal photoelectron spectrum, except that instead of a doublet state for each configuration there will be both a triplet and singlet state each time. If the neutral molecule starts with an open shell, as in O_2 or NO, more complications will arise from the unpaired core electron's spin interactions. We can expect the splitting in the spin multiplets for each configuration to depend mainly on the orbital overlaps, so splittings should become very small in large molecules where the final unpaired electrons can be far apart in space. As usual, these general considerations are more or less borne out by detailed molecular electronic structure calculations [1, 2]. Another difference from photoelectron spectra is that the core–valence ionised states have very short lifetimes before they undergo Auger decay to produce triply ionised final states. The consequent lifetime broadening usually eliminates any chance of observing vibrational fine structure.

The first core–valence double ionisation spectra were measured using the TOF-PEPECO technique by Hikosaka et al. [3] for Ne atoms and N_2 molecules. In molecules, core electron ionisation from different atoms within a molecule generally gives different core–valence ionisation spectra, except where the atoms are symmetry-equivalent, as in N_2. Where non-equivalent atoms of the same element are present, multiple slightly different core–valence spectra combine to form a composite form that may sometimes be decomposed theoretically into its components.

In the following sections we present examples of core–valence double ionisation spectra to illustrate the general themes mentioned here. Several of the spectra are compared with 'model' or 'synthetic' photoelectron spectra, created by taking the widths and intensities of bands from published photoelectron spectra and representing them in the form of Gaussians. The central band energies are shifted on a common scale to make the first peak match the first peak in the corresponding core–valence spectrum.

References

[1] H. Ågren and H.J.A. Jensen, *Chem. Phys.*, **172** (1993) 45.
[2] H.D. Schulte, L.S. Cederbaum and F. Tarantelli, *J. Chem. Phys.*, **105** (1996) 11108.
[3] Y. Hikosaka, *Phys. Rev. Lett.*, **97** (2006) 053003.

7.3 Core–valence spectrum of ammonia, NH_3

In ammonia and other molecules with only one atom besides hydrogen, an unpaired core electron and an unpaired valence electron can never be far apart, so singlet-triplet splitting is expected to be significant. This is borne out by the experimental core–valence spectrum in Fig. 7.3.1 and by detailed calculations.

On the basis of the electron configuration $1s^2 2s^2 1e^4 3a_1^2$ we can expect the lowest energy core–valence band to arise from $1s^{-1}3a_1^{-1}$ ionisation, which is one electron from the core and one from the N lone pair. The spectrum unmistakably shows two bands of similar width with peaks near 24 and 27 eV, which must be the triplet and singlet states from the resulting configuration. This identification, the energies of the bands and the splitting between them of 2.8 eV are all well reproduced by theory [2]. The main higher band with a peak at about 31 eV seems to correspond almost exactly to the positions of the 1e band in the photoelectron spectrum and we are tempted to assign it as such, but theory indicates that there is also a contribution from 3-hole-one particle configurations at the same energy [2]. The inner valence ($N2s^{-1}$) ionisation band in the photoelectron spectrum is only very weakly present, if at all, in the core–valence spectrum. The reason for this low intensity is not known to us at present.

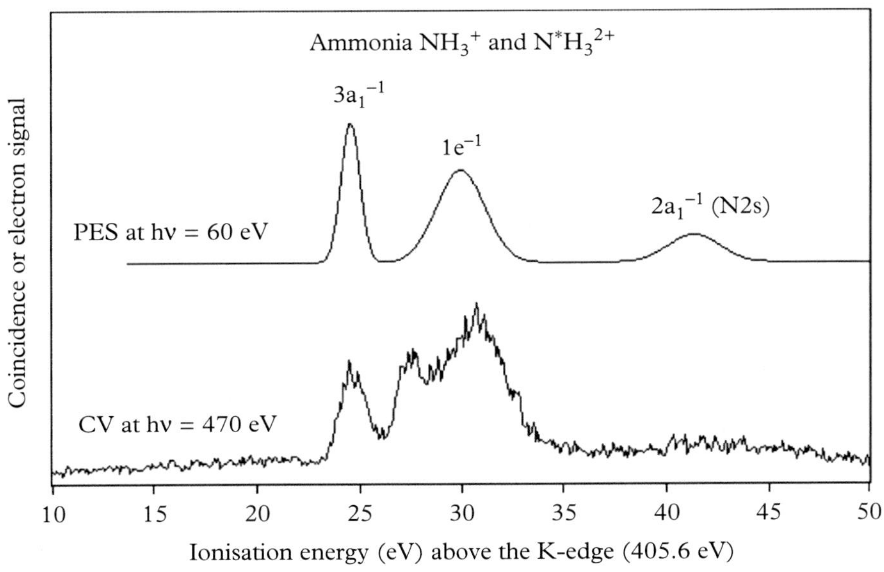

Figure 7.3.1 *Core–valence (CV) double ionisation spectrum of ammonia taken at 470 eV compared with the photoelectron spectrum (PES) at 60 eV, redrawn broadly from Banna et al. [1] and shifted to make the first bands coincide in energy.*

The simplest model for the energies of CV states, ignoring the distinction between singlets and triplets, is to set their energy as

$$E\,(CV) = IE\,(C) + IE_n + e^2/r_{12}$$

where IE_n represents the *n*th ionisation potential in the photoelectron spectrum and r_{12} is the separation of the two charges. The last two terms on the right correspond to the horizontal axis scale in Fig. 7.3.1 and the following figures. If we take the CV state energy in the lowest CV configuration of ammonia as 25 eV, this makes the Coulomb energy about 14 eV (25 − 11 eV) and the average distance between the positive charges about 1 Å, close to the conventional atomic radius of N at 0.7 Å. The agreement in this case may be fortuitous, but we shall see that the very simple model provides useful insights in other cases too. The core–valence spectra of methane [3] and water [4] exhibit the same characteristics as the core–valence spectrum of ammonia presented here. The first bands in those CV spectra are about 24 and 28 eV above the relevant core edges (Coulomb energies 10 and 15 eV), suggesting similar charge separation distances to that in ammonia.

References

[1] M.S. Banna, H. Kossmann and V. Schmidt, *Chem. Phys.*, **114** (1987) 157.
[2] R.R. Valiev et al., Mol. Phys., 115 (2017) 252.
[3] J.H.D. Eland et al., *Chem. Phys.* Lett., **485** (2010) 21.
[4] M. Mucke et al., *Chem. Phys.* Lett., **558** (2013) 82.

7.4 Core–valence spectra of carbon monoxide, CO

In a diatomic molecule like CO with two first row atoms we might expect that the charges, and thus the unpaired electrons in a core–valence ionised state, to be held about one bond distance apart by Coulomb repulsion, so reducing the splitting between singlets and triplets. The spectrum of CO teaches us that the actual splitting depends crucially on the location of the valence orbitals, as well as the location of the core orbital. The configuration of the neutral molecule can be written

$$1\sigma(\mathrm{O1s})2\sigma(\mathrm{C1s})3\sigma(\mathrm{O2s})4\sigma1\pi5\sigma \ldots\ldots\ldots\ldots\ldots\ldots\ldots {}^1\Sigma^+$$

with all orbitals doubly occupied. The 5σ orbital is characterised as a lone pair extending out from the carbon atom, located almost entirely on that atom. The 1π orbital and the 4σ orbital are located more strongly on O than on C [1]. The core–valence double ionisation spectra from carbon and oxygen 1s ionisations are quite different, as Fig. 7.4.1 shows.

Core–valence spectra of CO were first measured by Hikosaka et al. [2] at 410 eV for C1s and at 665 eV photon energy for the O1s-based spectrum. The spectra shown in

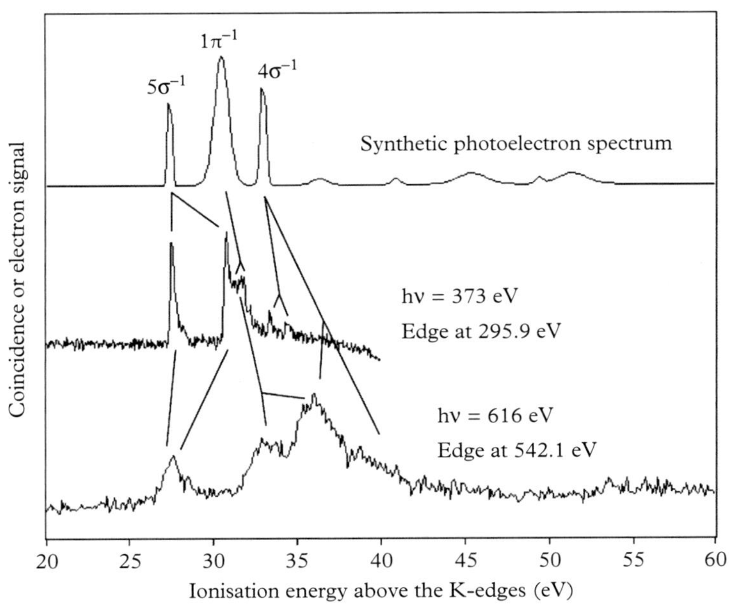

Figure 7.4.1 *Core–valence spectra of CO at photon energies of 373 and 616 eV with a model of the photoelectron spectrum as it may appear at about 70 eV. Lines are drawn to show how the valence orbitals give birth to triplet and singlet states in the dications, following the analysis of Hikosaka et al. [2].*

Fig. 7.4.1 have been measured at lower photon energies, 373 and 616 eV, to permit better electron energy resolution but are otherwise the same. Lines in the figure indicate how, according to theory [2, 3], ionisation from each valence orbital accompanied by a core hole produces splitting into a triplet at a lower energy and a singlet state at higher energy in each case. The 5σ valence ionisation for instance shows strong splitting with a C1s hole, but no measurable splitting with an O1s hole. Similarly, the 1π valence ionisation shows wider splitting with the O1s hole than with the C1s hole, in line with its localisation more on oxygen. Full theory tells us that this sort of analysis is an oversimplification, of course, because in reality the ionic states have mixed configurations and are not just simple two-hole states.

In N_2 and O_2, by contrast to CO, there is no localisation of valence orbitals on particular atoms, but the core–valence spectrum of N_2 [4] still shows some, but much less variation in singlet-triplet splittings in ionisation from different valence orbitals. Because of the open shells in the neutral molecules, the core–valence spectra of NO and O_2 [5] are complicated by the possible formation of states with more than two different spin multiplicities. As neutral NO has only one electron in its outermost orbital, the lowest energy core–valence ionised state is unsplit and appears as a single sharp band in both N- and O-based CV spectra [6]; the higher energy core–valence states of NO^{2+} have not yet been analysed. In the core–valence spectrum of O_2, for which a full theoretical analysis has been done to accompany the experimental spectrum [5], the lowest state has just one electron remaining in the antibonding π_g orbital to couple to the 1s core hole. The two unpaired spins give a triplet and a singlet, with a calculated splitting of 0.56 eV, which is less than the experimental width of the first band. When ionisation is from any of the deeper valence orbitals, quintet states are formed too, greatly complicating assignment of the spectral features.

References

[1] J.N. Murrell, S.F.A. Kettle and J.M. Tedder, *Valence Theory*, John Wiley, London, 1965.
[2] Y. Hikosaka et al., *J. Chem. Phys.*, **127** (2007) 044305.
[3] Y. Valiev et al., *Mol. Phys.*, **115** (2017) 252.
[4] Y. Hikosaka et al., *Phys. Rev. Lett.*, **97** (2006) 053003.
[5] E. Andersson et al., *Phys. Rev. A*, 78 (2008) 023409.
[6] Unpublished work of the authors' group.

7.5 Core–valence spectra of CO_2, OCS and CS_2

The core–valence spectrum of CO_2 based on removal of a C1s electron quite closely resembles the shifted valence photoelectron spectrum, as Fig. 7.5.1 shows.

The similarity between the $C1s^{-1}V^{-1}$ core–valence spectrum of CO_2 and the normal valence photoelectron spectrum comes in part from the smaller singlet-triplet splitting to be expected in larger molecules, particularly where electrons are delocalised. The first band at 24.8 eV corresponds to removal of an electron from the π_g orbital, which has no density at the C atom, so the splitting is particularly small. This is also the situation in the related case of CS_2 [3], where the $C1s^{-1}V^{-1}$ core–valence spectrum has been analysed theoretically. The next valence orbital, π_u, has maximum density at the C atom, so a wide singlet-triplet splitting is expected and has been confirmed in the CS_2 spectrum [3]. The first part of the second band in the CK core–valence spectrum of CO_2 must be the triplet from $C1s^{-1}\pi_u^{-1}$, while the singlet is hidden in the higher-energy part of the band. Core–valence ionisations involving the two σ-orbitals show only minor splitting in CS_2 [3], and the same may be assumed to be true in CO_2. When we turn to the $O1s^{-1}V^{-1}$ core–valence spectra the situation is quite different. The first band must involve the original π_g orbital which has its density all on the O atoms, so a significant singlet-triplet splitting is expected. The first band at 26.4 eV and the second at 30.9 eV may be the triplet and singlet from this $C1s^{-1}\pi_g{}^{-1}$ ionisation, but this remains to be confirmed. What is clear is that the core–valence bands are at higher energies in $O1s^{-1}V^{-1}$ than in $C1s^{-1}V^{-1}$ ionisation, suggesting that the charges are closer together in the doubly ionised states. Perhaps more significantly, when the core electron is removed from one oxygen atom

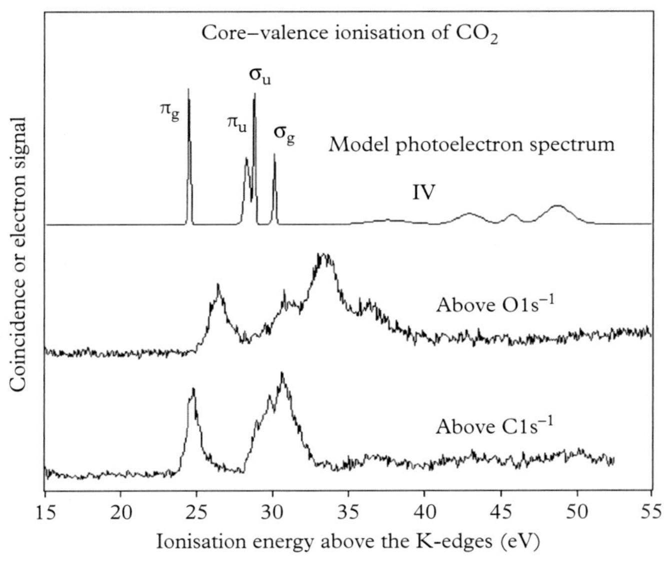

Figure 7.5.1 *Core–valence spectra of CO_2 above the two 1s ionisation (K) edges whose energies are noted in the figure, at photon energies of 360 eV for the C edge and 603 eV for the O edge, with a model photoelectron spectrum based on the spectrum in ref. [1] for the valence bands and on the spectrum in ref. [2] for the inner valence bands.*

the symmetry is reduced to $C_{\infty v}$, so there is no longer any g/u distinction. The spectrum might be expected to resemble that of a hypothetical linear [OCF^+] ion, as the removal of a 1s electron is similar in its effect to increasing the nuclear charge by one unit. This sort of effect has been examined explicitly in the related case of SO_2 core–valence ionisation [4], where it is shown to enforce localisation of the core charge on one O atom, despite the original indistinguishability of the two outer atoms.

In the closely related molecule OCS, the two outer atoms are already non-identical in the neutral molecule. The three core–valence spectra with holes in the C1s, O1s and S2p orbitals have been measured and compared with calculations using several different theoretical methods [5]. The computed singlet-triplet splittings are all small and are apparently unresolved in the spectra. The spectra based on $O1s^{-1}V^{-1}$ ionisation and $C1s^{-1}V^{-1}$ ionisation (Fig. 7.5.2) bear very much the same relation to the photoelectron spectrum as do the equivalent spectra for CO_2. The $S2p^{-1}V^{-1}$ core–valence ionisation spectra of OCS are complicated by the presence of the two edges (L_3 and L_2) where $^2P_{3/2}$ and $^2P_{1/2}$ holes are created. When two copies of the photoelectron spectrum are summed together with the same spacing in energy as the edges (ca 1.2 eV) and with the statistical 2:1 intensity ratio, the sum resembles the CV spectra quite closely [4]. This is seen clearly in the low energy bands of CS_2 [5], illustrated in Fig. 7.5.3.

The first bands in the CV spectrum of CS_2, both from $S2p^{-1}\pi_g^{-1}$ ionisation, are split by the spin-orbit energy separation of the edge (1.3 eV) and reproduce the 2:1 ratio of statistical weights in their intensity pattern. This doubling must presumably persist in the whole CV spectrum, but is less clear at higher energies because of congestion. In contrast to Fig. 2 of ref. [5], where a model photoelectron spectrum based on band intensities at $h\nu = 21.22$ eV was used for comparison with the CV spectrum, we here compare the experimental CV spectrum with a real photoelectron spectrum measured at $h\nu = 40.81$ eV where the energy spacings are the same but the relative intensities

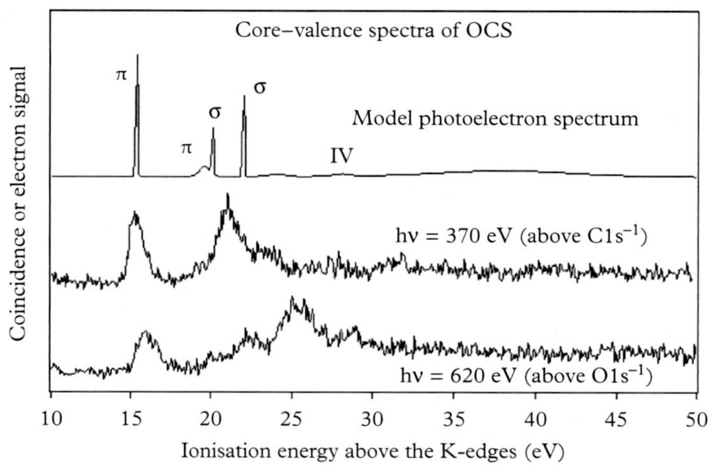

Figure 7.5.2 *Core–valence double ionisation spectra of OCS above the C1s and O1s edges compared with a model photoelectron spectrum based on ref. [1] for the valence bands and on ref. [6] for the inner valence bands.*

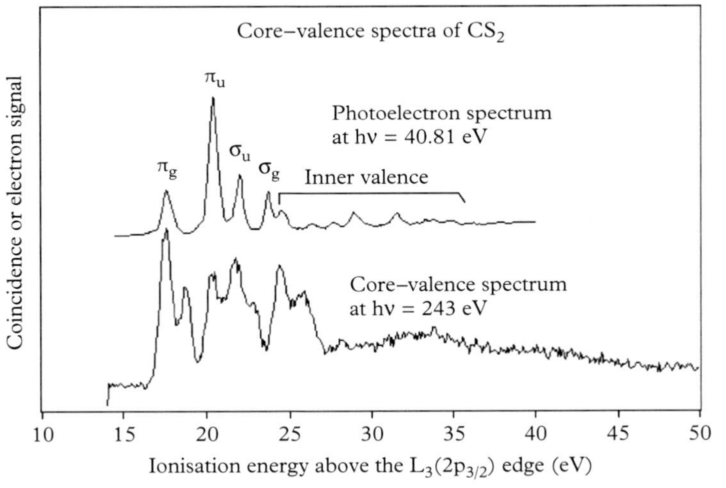

Figure 7.5.3 *Core–valence spectrum of CS_2 on a scale of energy above the L_3 edge (170.8 eV) compared with a shifted photoelectron spectrum taken at 40.81 eV photon energy.*

are notably different. The photoelectron spectrum in Fig. 7.5.3 closely resembles the spectra reported by Roy et al. [7] at similar energies, and includes the inner valence bands which the model spectrum did not. It is apparent that in ionisation from the π_u orbital, relative intensity is much greater in the 40.81 eV photoelectron spectrum than in the CV spectrum. Relative band intensities in the CV spectrum hardly vary at all as photon energy is varied from 220 eV to 243 eV [3], but photoelectron band intensities change significantly between $h\nu = 21.22$ eV and 40.81 eV [7]. This difference may be related to the sensitivity of photoelectron band strengths to shape resonances in the one-electron continuum. It illustrates an aspect of the current situation, where spectral intensities, in contrast to state energies, are hardly ever calculated theoretically.

A second conundrum brought out by Fig. 7.5.3 is the peculiarity of inner valence bands in the CV spectrum. Whereas in the photoelectron spectrum they are quite numerous and clearly structured, in the CV spectrum they are very broad and apparently fewer in number. The reason for this difference, which is also apparent in the CV spectra of CO and CO_2, is entirely unknown at present.

References

[1] A.W. Potts and G.H. Fatahallah, *J. Phys. B: At. Mol. Opt. Phys.*, **13** (1980) 2545.
[2] H.-J. Freund, H. Kossmann and V. Schmidt, *Chem. Phys. Lett.*, **123** (1986) 463.
[3] E. Andersson et al., *J. Chem. Phys.*, **133** (2010) 094305.
[4] J. Niskanen et al., *Phys. Rev. A*, **85** (2012) 023408.
[5] J. Niskanen et al., *Phys. Rev. A*, **82** (2010) 043436.
[6] D.M.P. Holland and M.A. MacDonald, *Chem. Phys.*, **144** (1990) 279.
[7] P. Roy et al., *J. Chem. Phys.*, **87** (1987) 2536.

7.6 Core–valence spectra of CF_4

In carbon tetrafluoride, the outer valence orbitals are almost all based on the fluorine atom, mostly as 'lone pairs'. The core of the C atom is far away, which may explain the very close match between the photoelectron spectrum (shifted up by 8.8 eV) and the $C1s^{-1}V^{-1}$ core–valence spectrum, as seen in Fig. 7.6.1.

It is notable in Fig. 7.6.1 that the $3t_2$ and $4a_1$ orbital-based states, as seen in the $C1s^{-1}V^{-1}$ spectrum, are displaced to higher relative energy than the states derived by ionisation of the outer orbitals. This is presumably because $3t_2$ and $4a_1$ have some C character in their makeup, while the outer valence orbitals are purely fluorine atom combinations. The $3t_2$ core–valence band seems to be split, possibly into a singlet and triplet, which would also indicate close unpaired electron localisation. When an electron is removed from an F1s orbital by contrast, the spectrum becomes much more spread out and hard to relate to the photoelectron spectrum. This is likely due in part to the symmetry breakdown and consequent Jahn–Teller distortion caused by core ionisation at one F atom. On the other hand, the lowest CV ionisation is no higher above the K-edge than when the core electron is removed from the C atom; presumably the valence charge is well spread over all the outer atoms. Very similar contrast between core ionisation at a central atom and core ionisation at an outer atom is seen in the CV spectra of CCl_4 and SF_6 [2].

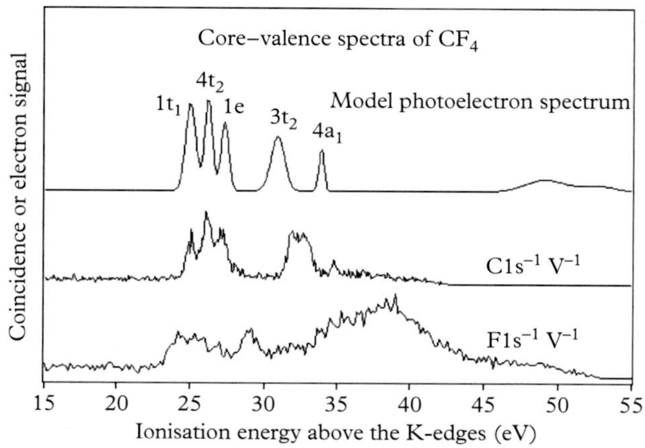

Figure 7.6.1 *Core–valence spectra of CF_4. The C1s edge is at 301.8 eV and the spectrum was taken at 375 eV; the F1s edge is at 696.5 eV and the spectrum was taken at 770 eV. The photoelectron spectrum is derived from data in ref. [1], with arbitrary relative intensity for the deep inner valence states.*

References

[1] G. Bieri, L. Åsbrink and W. von Niessen, *J. Electron Spectrosc. Related Phenom.*, **23** (1981) 281.
[2] Unpublished work of the authors' group.

7.7 Core–valence spectra of Si(CH$_3$)$_4$

Tetramethylsilane (TMS) provides another example of CV ionisation of a molecule with contrast between the central atom and outer atoms, but here the difference in spectra is extreme and may be informative.

The extreme contrast between the two core–valence spectra in Fig. 7.7.1 resides mainly in the intensities of the different bands. For the first band, whose valence part must be $3t_2^{-1}$ ionisation, the weakness when accompanied by a C1s hole and strength when accompanied by a Si2p hole may be related to the orbital composition, as this is the valence orbital with most Si atomic character [2]. The two lower-energy orbitals in the broad central band of the photoelectron spectrum are mainly C—H bonding, so the very strong band in the C1s^{-1}V^{-1} spectrum may possibly be ascribed to them, while $2a_1$ also has some Si character. Beyond this we cannot get further without theoretical aid.

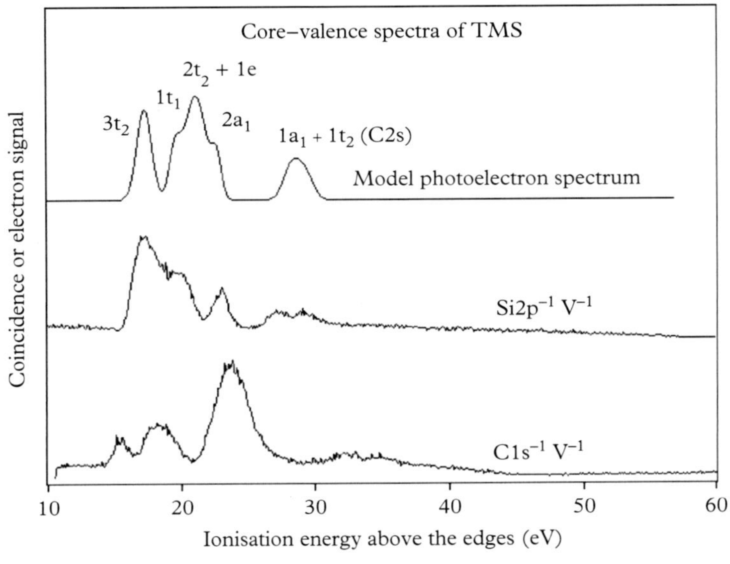

Figure 7.7.1 *Core–valence spectra of TMS taken at 360 eV for the C1s edge and 170 eV for the S2p edges and compared with a model photoelectron spectrum based on ref. [1]. The C1s edge position was measured as 289.5 eV and the Si2p$_{3/2}$ edge position as 106.1 eV. The spin–orbit splitting in Si2p is about 0.5 eV, too small to affect the CV spectrum significantly.*

References

[1] J.E. Bice et al., *Inorg. Chem.*, **26** (1987) 4106.
[2] D.G.J. Sutherland et al., *Organometallics*, **13** (1994) 3671.

7.8 Core–valence spectrum of C_6H_6

The benzene molecule is so large, and its orbitals are so delocalised that we can expect only small singlet-triplet splittings. This is borne out by the comparison between the CV spectrum and the photoelectron spectrum in Fig. 7.8.1.

The similarity between the CV spectrum and the normal photoelectron spectrum in Fig. 7.8.1 is so close that there is no apparent need for calculations to enable a first interpretation. The lowest energy band in the CV spectrum corresponding to $C1s^{-1}1e_{1g}^{-1}$ ionisation is split into a doublet with a spacing of 0.9 eV, which would be surprisingly large as spin-splitting into a triplet and singlet. Instead we believe that it demonstrates the reduction in molecular symmetry from D_{6h} to C_{2v} on ionisation of the 1s core on a single carbon atom. The effect of this is to convert benzene into something closely resembling pyridine in molecular orbital structure, as $1s^{-1}$ ionisation increases the nuclear charge seen by the outer shell of that particular atom by +1. In pyridine, benzene's degenerate $1e_{1g}$ orbital is split into two distinct π-orbitals with a spacing of 0.75 eV [1]. All the CV configurations derived from originally degenerate (e-symmetry) orbitals will be split in a similar way, to differing extents, and each resulting configuration can be split further by spin-spin and other lesser interactions.

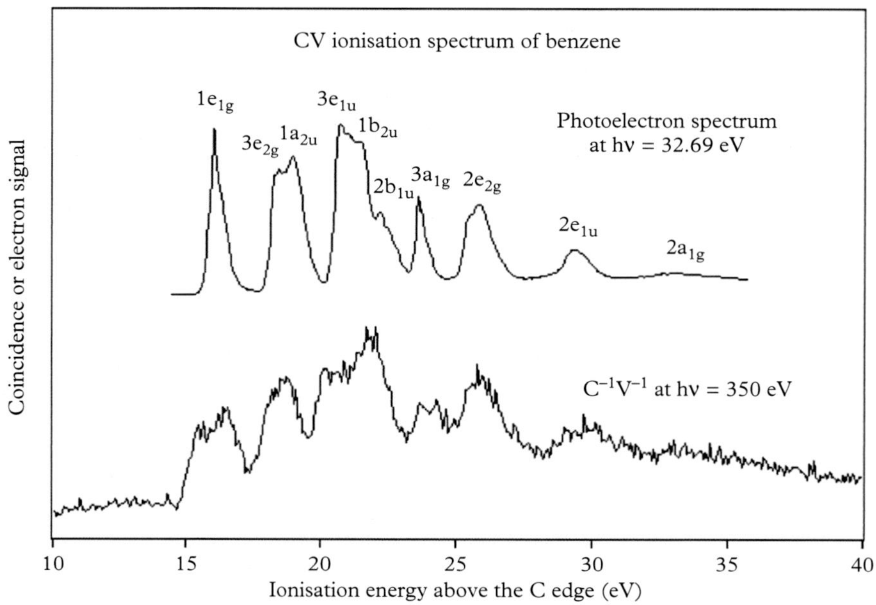

Figure 7.8.1 *Core–valence spectrum of benzene compared with a photoelectron spectrum taken with the TOF-PEPECO apparatus using NeIII light and shifted up in energy by 6.75 eV. The 1s edge of benzene is at 290.3 eV.*

Most of these splittings are too small to be visible in the experimental CV spectrum, but must seriously complicate a full theoretical description.

To match the CV spectrum, the photoelectron spectrum had to be shifted up by just 6.75 eV. According to the very simple model mentioned in the introductory section within this chapter (Section 7.2), such a shift corresponds to an intercharge distance of 2.13 Å, which is only a little less than the mean diameter of the benzene ring (2.8 Å). This agreement encourages us to think that for molecules of this size and larger, the simple model may have useful predictive power for CV state energies.

Reference

[1] R. Gleiter, V. Hornung and E. Heilbronner, *Helv. Chim. Acta*, 55 (1972) 255.

7.9 Core–valence spectrum of fullerene, C_{60}

The C_{60} molecule has icosahedral symmetry and is almost a sphere of 10 Å diameter. The outermost π-orbital, h_u, is fivefold degenerate and contains 10 electrons. In such a large molecule both the singlet-triplet splitting and the effect of symmetry reduction by core ionisation at a single C atom are probably small. In this case we have no nitrogen-containing molecule to guide us, and as yet no theoretical calculation of the core–valence spectrum. In the experimental CV spectrum, Fig. 7.9.1, the first band is not fully resolved. A good fit between the CV spectrum and the first two bands in the photoelectron spectrum is found if the first feature, a shoulder 11.1 eV above the C1s edge (290.1 eV), is identified as $C1s^{-1}h_u^{-1}$ ionisation. A reasonable, but less close fit is also found, however, if the shoulder is ignored and the peak of the first band is assigned as this transition. As spectral congestion is severe in both spectra, we cannot decide on the correct interpretation at this time.

The photoelectron spectrum must be shifted in energy by 2.1 eV in order for its first peak to match the first peak in the CV spectrum. This is the smallest shift for any molecule examined so far. According to the simplest Coulomb model, this would indicate that in the doubly charged ion the charges are about 7 Å apart, in keeping with the size of the molecule.

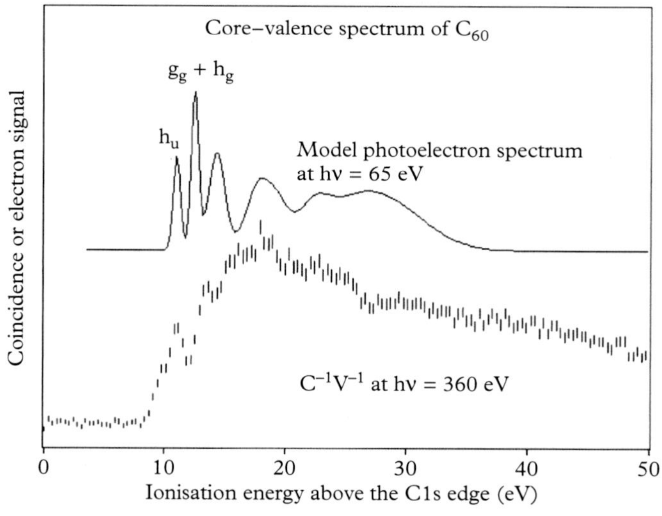

Figure 7.9.1 *Core–valence spectrum of C_{60} compared with a model photoelectron spectrum of the molecule based on the 65 eV gas phase measurement of Liebsch et al. [1].*

Reference

[1] T. Liebsch et al., *Phys. Rev. A*, 52 (1995) 457.

Appendix 1
Mechanisms of double photoionisation

The fundamental distinction between direct and indirect photoionisation pathways has been mentioned many times in this book. The direct process takes place in a single two-electron step and can be written symbolically as

$$m + h\nu \rightarrow m^{2+} + 2e^{-}.$$

An indirect process involves at least two steps; firstly a singly charged intermediate state is formed by a regular one-electron dipole-allowed transition, then in the simplest indirect process that intermediate state spontaneously emits a second electron in a monopole transition:

$$m + h\nu \rightarrow m^{+} + e_1^{-};$$
$$m^{+} \rightarrow m^{2+} + e_2^{-}.$$

For atoms or molecules that possess no singly charged states above the double ionisation limit, only direct double ionisation is possible; this is the case for H^-, He, H_2, Li^+, Be^{2+} and all other He-like species. In complex atoms and molecules both direct and indirect double ionisation processes are possible, and as the lifetime of an intermediate state gets shorter, the distinction eventually becomes blurred. For each single process, the uncertainty ΔE_2 in the energy of electron e_2^- is related to the lifetime τ by the usual uncertainty relationship $\tau\Delta E_2 \leq h/2\pi$. If the energy uncertainty ΔE_2 is of the same magnitude or bigger than the energy E_2 itself, the distinction between direct and indirect pathways is lost. Because the formation of an intermediate singly ionised state involves only single electron ejection and is always dipole-allowed, indirect ionisation is usually dominant over the direct process. This is most notably the case in the normal Auger effect, where the intermediate state is formed by simple electron ejection from an occupied orbital. In cases where formation of a superexcited intermediate state requires the participation of two electrons, usually the excitation of one and ejection of another, the intensity advantage of an individual indirect process over the direct process is much less marked. But there are generally many possible intermediates of this two-electron type, so overall the indirect processes tend to remain dominant. Because of the way the cross-sections vary with photon energy, roughly constant above threshold for indirect processes (like single ionisation) but rising with increasing energy for direct double ionisation, indirect processes tend to be dominant for low-photon energies but diminish in importance as photon energies are higher.

Despite the concentration of this book on molecules rather than atoms, we shall discuss direct double ionisation in atoms at some length because of the great fundamental interest and importance of the direct process for both atoms and molecules.

Direct double photoionisation

Almost all papers on direct double photoionisation until recently advertised their importance and relevance on the grounds that the process itself is forbidden in simple theory, and becomes allowed only by electron correlation, a hot topic. Indeed, it is still accepted that apart from very exceptional circumstances, one photon interacts directly with just a single electron. Electron correlation can be understood as having to come in after this initial step, and is made explicit in the classical Wannier model of double electron escape near threshold [1]. If two electrons are to both get away from a residual positive ion with only a small excess energy to share between them, they must leave maintaining equal distances from the ion and ideally with a 180 degree angle between their trajectories. If one electron gets further away from the ion than the other it sees a more screened positive charge and escapes faster, while the other electron sees a less screened charge, slows down, and is recaptured. As the available energy is increased, more and more different trajectories lead to successful escape, and the importance of this correlation diminishes. The effect of the need for correlation between trajectories in the final state is to change the way the cross-section σ for double ejection varies with excess energy E_X above threshold, from simple proportionality, $\sigma = kE_X$ to a power law, $\sigma = kE_X^m$ with an exponent m of 1.056 in double photoionisation. It also sets requirements on the angular distribution of the electrons relative to one another, which depend in a rather complex way on the symmetry of the ion produced [2]. The theory leads to a further prediction that in the energy range where it is valid the distribution of energy shared between the two electrons should be flat. These predictions are all quite well confirmed by experiments on atomic He and by quantum calculations, but there is also a quantum mechanical model [3] which makes predictions that differ in some important details. No experiments have yet ruled decisively in favour of either the classical or the rival quantum mechanical theory. There is no full consensus on the extent of the 'threshold' range where either theory can expect to hold, but experiments suggest a range of a few eV.

These theories about escape in the final state do not tell us anything about how a single photon brings about the ejection of two electrons in the first place. There are two model mechanisms supported by theory and by experiment, now called *knock-out* (KO) and *shake-off* (SO). In the KO mechanism, the photon interacts with one electron, tending to eject it; that primary electron then interacts with a second electron in an inelastic collision, causing it to be ejected too. The existence of the KO mechanism was first predicated experimentally by Samson and coworkers [4, 5] who noticed that some double photoionisation yields vary as a function of photon energy in the same way as the cross-sections for electron impact ionisation of ions. The SO mechanism has long been familiar as an explanation of features in electron spectra induced by X-ray absorption and by radioactive atom decay. Its basis is that the removal of one electron, suddenly increasing the charge, gives such a shock to the system that electrons can be excited to higher orbitals (shake-up) or ejected (shake-off, SO). Sophisticated theoretical methods have since been developed to describe the whole process of direct double photoionisation in a fully quantum mechanical way, and support a model in which both KO and SO mechanisms operate at all energies to different extents. In the low energy (threshold) region KO is dominant, but it starts to fade out for He at excess energies of a few hundred eV. The SO mechanism makes a negligible contribution at threshold but becomes more and more important as the excess energy increases; its intensity saturates and becomes constant above a characteristic energy of the primary electron [6]. Because both mechanisms have the same starting point and end point there can be interference effects between them, but theoretical models using an incoherent sum of the two have given a satisfactory explanation of experimental results up to now [7]. Although the connection may seem tenuous, the SO and KO mechanisms can also be considered as embodiments of the electron correlation omitted from the simplest theory.

Using the TOF-PEPECO method, one obtains data on the distribution of excess energy between the two electrons of an ejected pair for each and every dication state and photon energy in the course of measuring the double ionisation spectra. Where ionisation mechanisms are discussed these are also called singly differential cross-sections (SDCS); their shape expresses the balance between SO and KO in direct double photoionisation and often betrays the presence of indirect pathways. For a purely direct process the distribution must be a smooth monotonic function of the electron energy. For low excess energies the distributions are rather flat, but as photon energy and excess energy increase the distributions take on a hollow 'U' shape, concentrating intensity on one fast and one slow electron. The degree of hollowness depends on the relative proportions of SO and KO contributing mechanisms, which are functions of the excess energy available. Indirect pathways can introduce symmetrical pairs of peaks corresponding to each resolved intermediate state. Unfortunately the peaks due to intermediate states often overlap and cluster at the lowest and corresponding highest electron energies; this can make them hard to recognise. Figure A.1.1

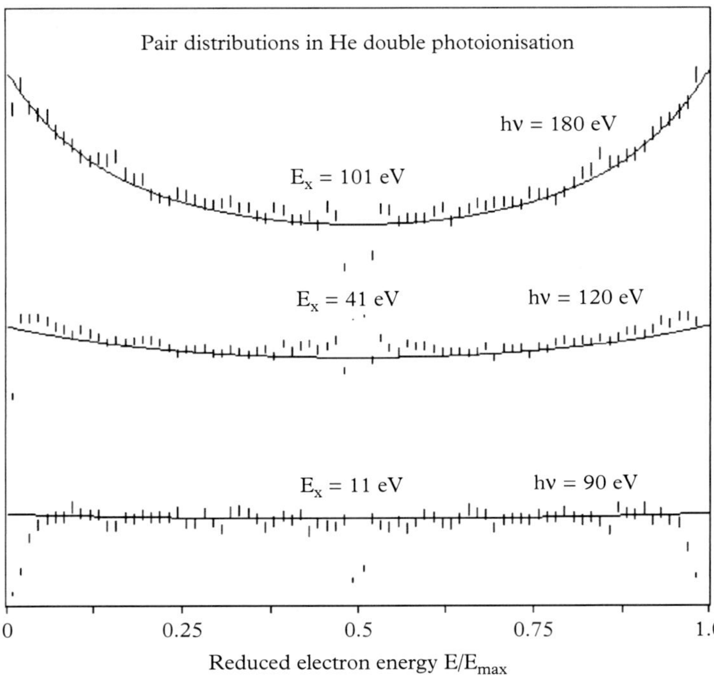

Figure A.1.1 *Electron pair distributions in double photoionisation of He at photon energies of 90, 130 and 180 eV, corresponding to the excess energies E_X shown in the figure and illustrating the increasing hollowness of the distributions at higher photon energies. The data are shown on a scale of reduced electron energy (E/E_{max}, $E_{max} = h\nu - DIE$) and represented by 2σ error bars. Solid lines show the shapes of an arbitrary function which fits all the distributions up to an excess energy of 200 eV with E_X as the only input parameter.*

shows examples of pair distributions in pure direct double photoionisation, using He as the example. In fact, the only experimentally convenient neutral species for which we can be sure that only a direct process occurs are He and H_2.

Indirect double photoionisation

From an experimental point of view, we classify indirect pathways as follows.

1. Classical Auger decay after ionisation from an inner shell localised on a single atom.
2. Auger decay of an inner valence or deep valence singly ionised state.
3. Autoionisation of 2h-1p states whose 2h cores are excited states of the dication.
4. Autoionisation of fragments from dissociation of singly charged intermediates.

Classical Auger decay

In the classical Auger effect a vacancy is first produced by photon impact or other means in an inner shell of one atom within a molecule. The hole state so produced persists for a short time, usually a few fs for deep inner shell holes, before the vacancy is filled by a valence or other less strongly bound electron located on or close to the same atom, releasing enough energy to cause ejection of a second electron, usually located close by. If the original molecule had a closed shell, the final dication produced is most likely to be in a singlet state. If the initial vacancy and the final double vacancies are all in orbitals of the same principal quantum number the process is particularly fast and is called a Coster–Kronig transition. The spectra of Auger electrons are spectra of final doubly ionised states and are different when the initial vacancy is on different atoms. TOF-PEPECO experiments have shown that the Auger spectra are different even when the first hole is created on chemically inequivalent atoms of the same element, such as C atoms, within a molecule [8, 9]. Because of the short lifetimes of deep inner shell holes and the consequent uncertainty broadening, classical (single electron) Auger spectra generally suffer from poor resolution. This problem is completely overcome in principle by the TOF-PEPECO technique and other coincidence techniques where the energies of both the photoelectron and the Auger electron are measured. The sum of the two energies or strictly $h\nu - (E_1+E_2)$ corresponds to the energy difference between two quantum states, the molecular ground state and a stationary or quasi-stationary state of the dication, so the core–hole lifetime broadening is eliminated. In practice, it is difficult to achieve this with a normal TOF-PEPECO setup because the Auger electron tends to have a very high energy, reducing the instrumental resolution. The answer is to have two spectrometers viewing the same ionisation events in coincidence, one optimised for high and one for low energies, but so far this has rarely been done. Within their short lifetimes, inner shell hole states can still evolve, and H atoms, because of their low mass, can move significantly. Their equilibrium conformations, particularly bond lengths, are generally different from those of the neutral molecules, so band shapes (Franck–Condon envelopes) in Auger spectra are different from those in low energy double photoionisation. In principle, X-ray fluorescence competes with Auger decay in disposing of the energy of inner shell hole states, but for first row atoms ionised in the 1s shell and hole states of similar or lower binding energy ($2p^{-1}$ in Si–Ar, $3d^{-1}$ in Se, Br and so on) the fraction of such fluorescence is negligible.

Auger decay from inner valence states

Auger decay from an inner valence or deep valence state is prevalent in the double ionisation of large molecules where the double ionisation threshold is low and/or vacancies can be created in states based on the 2s shells of first row atoms. States based on C2s are found in the region of 20–25 eV, from N2s around 35–40 eV, from O2s near 40 eV, and from F2s near 42 eV. The energies depend strongly on the chemical environment because the states involved interact with valence orbitals, and generally all these states tend to have heavily mixed configurations. It is difficult to isolate their Auger decays from other double photoionisation pathways, so there has been very little research in this area. We know this sort of process happens and is probably a major decay pathway for inner valence monocation states where it is energetically possible, but we know little more. It is not the only possible decay pathway, because molecular ion fragmentation may compete with electron ejection; the extent of such competition has not yet been quantified. Pair distributions dominated by Auger processes are illustrated in Fig. A.1.2, where they are compared

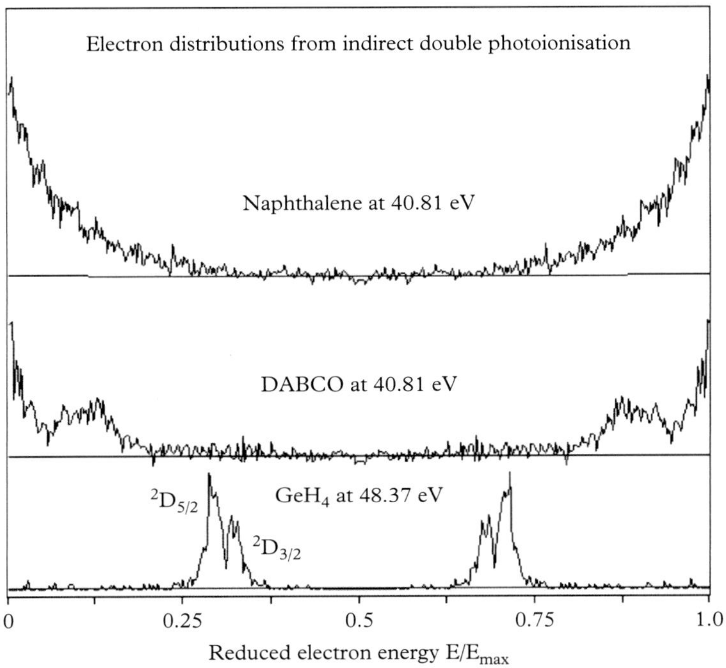

Figure A.1.2 *Electron pair distributions dominated by indirect double ionisation via core–hole states (GeH_4) and inner valence hole states. These are all distributions for production of the ground states of the molecular ions. Each experimental distribution has a solid line beneath it, showing the form of distribution for direct double photoionisation of He at the same excess energy (see text). DABCO is 1,4-diazabicyclo[2,2,2]octane, $C_6H_{12}N_2$, with two N atoms held rigidly apart by three $-C_2H_4-$ bridges. Naphthalene is a bicyclic aromatic hydrocarbon; its structure is shown in Fig. 5.7.1.*

with model direct distributions based on the hitherto untested hypothesis that excess energy is the only parameter affecting their shape, so they have the same form as in He ionisation.

Autoionisation of 2h-1p states

Autoionisation from 2h-1p states, often Rydberg states belonging to series converging on excited states of a dication, is very prominently visible in atomic double ionisation, as Fig. A.1.3(c) illustrates. In molecules the effect is undoubtedly present, but the signs of it are more subtle and difficult to discern, because instead of sharp atomic lines we have overlapping bands, each as wide as the 2h dication level to which they converge. There can even be series of Rydberg states independently converging on each resolved vibrational level within an electronic state band. Clear evidence for the effect in molecules would be a difference in electron pair distributions between formation of the lowest dication state or states, for which plentiful higher states capable of supporting convergent 2h-1p states exist, and the highest 2-h dication state which has no higher states above it. To make a valid comparison we should ideally compare the pair distributions for the same excess energy above each state. Unfortunately, real molecules have too many high energy dication states, stretching right up to the triple ionisation energy and beyond, for such a comparison to be

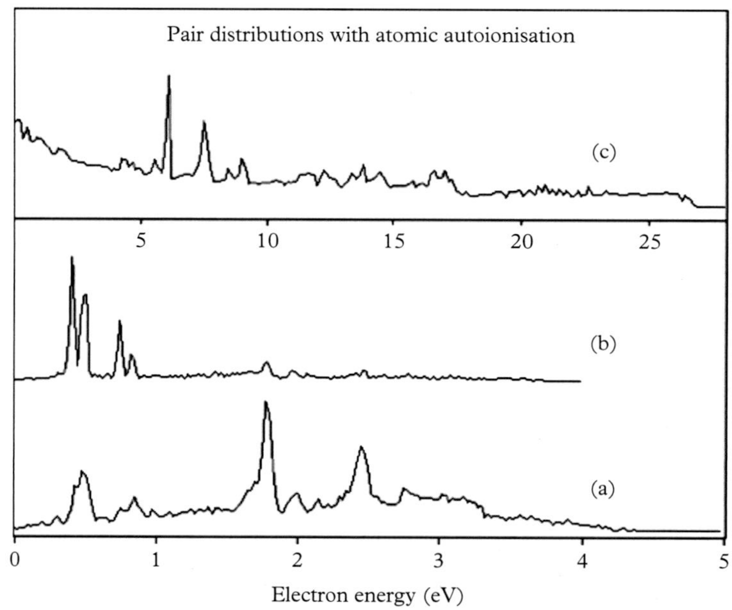

Figure A.1.3 *Lower energy halves of electron pair distributions from atomic autoionisation: (a) for formation of* $O^+ + NO^+$ *fragments from* NO_2 *at 40.81 eV; (b) for formation of* $O^+ + C^+$ *fragment ions from CO at 40.81 eV; and (c) for double ionisation to the* $^1D_2$ *level of* Ar^{2+} *at 48.37 eV, mainly via 2h-1p and more exotic superexcited states. In (a) and (b) all the peaks come from autoionisation of excited* O^* *atoms, giving the same electron energies but very different relative intensities and widths in the two cases.*

possible. The best we can do is to compare the pair distribution for the lowest state, which has most states above it, with a well-defined higher state which has fewer higher energy doubly ionised states or corresponding singly ionised 2h-1p states above it.

Autoionisation of fragments from molecular dissociation

The autoionisation of fragments from cation dissociation is most clearly exemplified by molecular oxygen, as mentioned in Chapter 2. The sequence of events can be written as

$$m + h\nu \rightarrow m^+ + e_1^-$$
$$m^+ \rightarrow a^+ + b^*$$
$$b^* \rightarrow b^+ + e_2^-.$$

This process is unambiguously recognised because the energies of the electrons from the final step are sharp lines, characteristic of the excited species b* (e.g. O*), known from their appearance in many different experimental contexts and also calculated from resolved atomic absorption spectra. Fragment autoionisation processes are seen in coincidence maps and pair distributions from double ionisation of several diatomic and triatomic species, but much less frequently for larger molecules. In most cases the autoionising fragments are atomic, particularly O*, N*, S*, I*, but there may be autoionisation of some molecular fragments too. It would be much more difficult to detect molecular fragment autoionisation, as the electron energies would be bands broadened by vibrational structure instead of sharp lines. As can be seen in the coincidence maps for CO, O_2, N_2O and NO_2 (Chapters 2 and 3), the atomic autoionisation appears over a wide range of double ionisation energies. This is because the autoionisation can occur at different separation distances, so more or less energy appears as kinetic energy of the fragments. The minimum distance to which the fragments must have separated before the instant of autoionisation can be inferred from the kinetic energy of the separating ion pair, since Coulomb repulsion will contribute an energy e^2/r_{12} in addition to any energy released in the initial dissociation. The fragments sometimes get as far as 20 Å apart before autoionisation.

Effects of resonant neutral states

If the photon energy used for double photoionisation matches the energy of a neutral state, double ionisation can occur by resonant double Auger decay as well as directly. At some resonances, particularly the π^* pre-edge resonances in unsaturated molecules, the cross-section for excitation is enormously enhanced relative to underlying ionisation continua. Single ionisation spectra produced by Auger decay from such resonances have been widely studied; they differ substantially from normal photoelectron spectra of ground-state molecules because the electron already excited to a virtual orbital in the resonance can act either as a *participator* or as a *spectator* in the Auger decay. Spectra produced in the participator process are similar to normal photoelectron spectra, but differ in detail because the equilibrium molecular structure in the excited state may differ from that in the neutral ground state. The spectator spectra, which retain one electron in a virtual orbital, are completely different and can be very complex. Double ionisation spectra produced by double Auger decay from resonances must be expected to show similar phenomena, with even more added complications from indirect pathways in the ionisation step. They have not yet been investigated in any detail.

In TOF-PEPECO spectra taken with HeII light, the chance that the photon energy accidentally matches a neutral resonance is small in most cases, because the photon energy is above all states derived from vacancies in valence orbitals. Even the inner valence orbital energies are well below 40 eV, except in molecules containing a fluorine or oxygen atom. For fluorine-containing molecules there is a serious danger of unanticipated indirect ionisation processes if 40.81 eV light is used, because inner valence hole states and 1h-1p neutral states based upon them are found at around this energy. With 48.37 eV or higher energy photons the danger is much reduced. For molecules containing only first row atoms from B to F, the photon energy range from 50 to 175 eV (or to 250 eV if B atoms are absent) is free of neutral states which could cause complications. For molecules containing heavy atoms of the later groups in the periodic table, particularly Ge to Br and Sn to I or atoms with full f-shells (W to Pb), there is a significant chance of hitting a pre-edge resonance at photon energies useful for double ionisation.

References

[1] G.H. Wannier, *Phys. Rev.*, **90** (1953) 817; *ibid.*, **100** (1955) 1180.
[2] F. Maulbetsch and J.S. Briggs, *Phys. Rev. Lett.*, **68** (2004) 1992.
[3] A. Tempkin, *J. Phys. B: At. Mol. Opt. Phys.*, 7 (1974) L450; *Phys. Rev. Lett.*, **49** (1982) 365.
[4] J.A.R. Samson, *Phys. Rev. Lett.*, **65** (1990) 2861.
[5] J.A.R. Samson, R.J. Bartley and Z.X. Hey, *Phys. Rev. A*, **46** (1992) 7277.
[6] T. Mukoyama, *X-ray Spectrom.*, **39** (2009) 142.
[7] T. Schneider, P.L. Chocian and J.-M. Rost, *Phys. Rev. Lett.*, **89** (2002) 073002.
[8] J.H.D. Eland et al., *Chem. Phys. Lett.*, **548** (2012) 90.
[9] H. Iwayama et al., *J. Chem. Phys.*, **138** (2013) 024306.

Appendix 2
Useful data for calibration

Calibration in TOF-PEPECO normally requires conversion from measured electron time of flight (TOF) to electron energy EE. In most cases, this can be done using the simple formula

$$EE = \{D/(TOF - t_0)\}^2 - E_0.$$

Here D, t_0 and E_0 are constants to be determined in a calibration procedure, representing the effective length of the flight path, a time offset and an energy offset, respectively. In practice D is fixed for a given apparatus, so only two constants must be determined afresh for each measurement or measurement period. The calibration normally starts with measurement of several spectra, usually single ionisation spectra of reference compounds containing sharp peaks of accurately known electron energy and ideally covering as wide a range as the data to be collected. The tables below list the data that has proved most useful to the authors for this task. The best values for the conversion constants can then be determined by an error minimisation procedure.

In some circumstances, or where the highest possible precision is desired, a more complex conversion formula may be needed, with more constants. This is generally necessary when electrons are retarded in the flight tube to improve energy resolution at the expense of a reduced range of energies covered. A polynomial expansion in powers of the inverse flight time (TOF – t_0, with trial values of t_0) is usually effective.

Most of the energies quoted here have been compiled from multiple sources and are presented without individual references to the original papers, as a guide to the practical process of spectrum calibration.

If the energy of the available photons is accurately known, that is, with atomic emission lines or energy-selected synchrotron radiation from a reliably calibrated monochromator, calibration can be effected using known ionisation energies. More complete data on atomic energy levels and ionisation energies are to be found in the NIST atomic spectra database [1] and in a recent book [2].

Energies of useful atomic lines and ionisation energies

Atomic emission lines

HeI: α; 21.218 eV. β; 23.085 eV. γ; 23.742 eV
HeII: α; 40.814 eV. β; 48.372 eV. γ; 51.017 eV
NeII: 26.814, 26.911 eV; 30.453, 30.549 eV (doublets)
NeIII: 32.687 eV (singlet)

Single ionisation energies (IEs) of helium to different He^+ states

Quantum number	IE (eV)
1	24.588
2	65.40
3	72.96
4	75.61
5	76.83

Single ionisation energies (eV) of the heavier rare gases

State	Ne	Ar	Kr	Xe
$^2P_{3/2}$	21.565	15.760	14.000	12.130
$^2P_{1/2}$	21.661	15.937	14.666	13.436
$^2S_{1/2}$	48.475	29.240	27.514	23.397
$^2D_{5/2}$			93.788	67.548
$^2D_{3/2}$			95.038	69.537
$^2P_{3/2}$		248.63		
$^2P_{1/2}$		250.78		
$^2S_{1/2}$	870.2			

If the $n\mathrm{d}^{-1}$ ionisations of Kr, Xe, the 1s ionisation of Ne or ionisations of other inner shells are used in this way, it is important to be aware of the effect of post-collision interaction (PCI). If the photon energy is at all close to the energy of the edge, the slow photoelectron may be overtaken by a faster Auger electron. This produces an energy exchange in which the photoelectron loses energy and the Auger electron gains it. To allow for this effect, the energy shift ΔE can be estimated using the formula

$$\Delta E = 1.884\,\Gamma\,\{1/\surd E_1 - 1/\surd E_2\}$$

where Γ is the intrinsic width of the hole states (ca 0.09 eV for Kr3d, 0.11 eV for $\mathrm{Xe4d}_{5/2}$, 0.14 eV for Ar2p, 0.27 eV for Ne1s) and E_1 and E_2 are the energies of the photoelectron and Auger electron in eV [3]. The photoelectron line *shape* is also strongly affected by post-collision interaction, so the estimated shifts should be used with caution.

Spectrum of molecular oxygen and its ionisation energies

Figure A.2.1 shows the photoelectron spectrum of molecular oxygen taken with the HeIα line, a spectrum which is particularly useful as it provides reference lines at low electron energy. The ionisation energies and electron energies of the most useful lines are quoted from Edqvist et al. [4] for the X and b states and from Baltzer et al. [5] for the B state. There are two more sharp lines, accessible using HeII light, at 24.564 and 24.756 eV ionisation energy, from the $c^4\Sigma_u^-$ state.

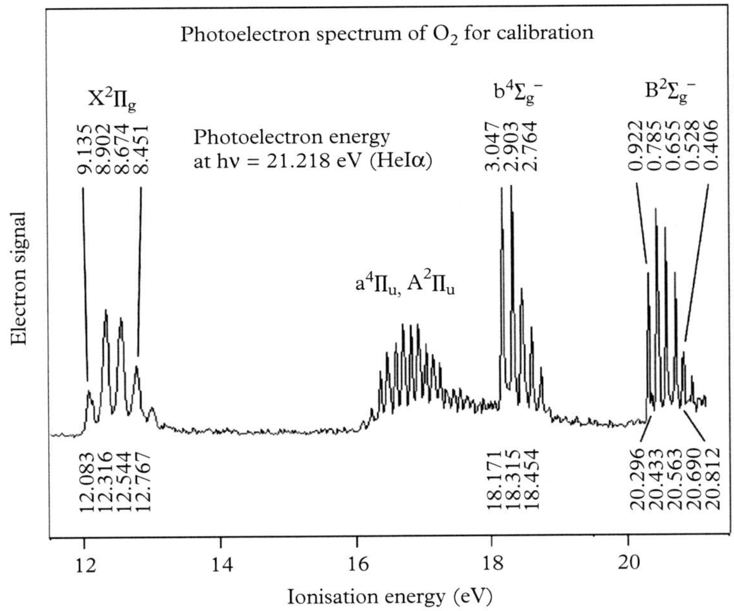

Figure A.2.1 *Photoelectron spectrum of oxygen measured using a 5.6 m magnetic bottle TOF apparatus under good conditions. Photoelectron energies of the lines which can most often be resolved with HeIα light are given above the spectrum and the ionisation energies are given below.*

Fixed electron energies

Strong lines in Auger spectra from $3d^{-1}$ ionisation in Kr and $4d^{-1}$ in Xe

Where the photon energy is only approximately known, as is sometimes unfortunately the case at synchrotron radiation facilities, or if only a broadband source of light is available, calibration must start with electron energies which are independent of the means of excitation. The Auger spectra from atomic gases match this requirement and the low energy Auger lines from Xe and Kr are the most useful for TOF calibration. With TOF-PEPECO apparatus and near-monochromatic light it is possible to isolate the Auger lines that come from each separate spin–orbit level of a multiplet hole state. This gives the clearest spectra, and is almost essential in the case of the Kr3d Auger spectrum, which is quite congested. The spectra in Figs. A.2.2 and A.2.3 provide the intensity patterns of the most intense lines, allowing experimental identification even if resolution is less than the ideal shown. The electron energies of the intense lines are listed later. The low energy Auger spectrum of Xe is produced by ionisation at any photon energy above 70 eV, while for Kr any energy above 100 eV will serve, but it is wise to choose photon energies where the photolines do not interfere with the Auger spectra.

Energies (eV) of the strong Xe Auger lines [6]

$^2D_{5/2}$: 8.300, 12.74, 14.169, 15.270, 19.686, 29.969, 32.323, 33.229

$^2D_{3/2}$: 10.279, 14.703, 16.146, 17.249, 21.665, 31.946, 34.302

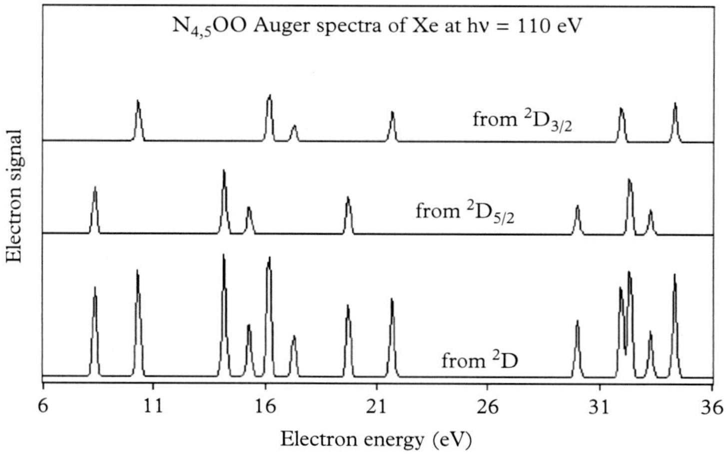

Figure A.2.2 *Strong lines in the 4d Auger spectrum of Xe, as excited at 110 eV photon energy, redrawn with idealised resolution of 0.2 eV from experimental TOF-PEPECO spectra. There are more weak lines in the full spectrum [6]. The photon energy is a convenient one where the photolines do not interfere with the Auger spectrum, but its exact value is unimportant.*

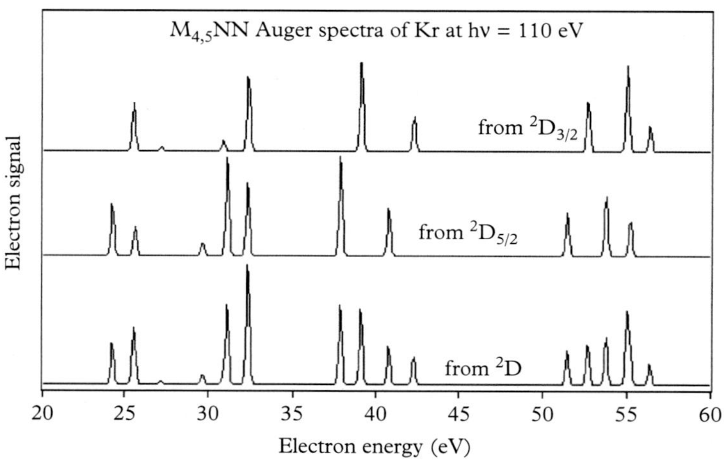

Figure A.2.3 *Schematic Auger spectra of Kr from $3d^{-1}$ ionisation. The line intensities and positions are from experimental TOF-PEPECO data while linewidth has been fixed as 0.2 eV. Weak lines are not included and in the full real spectra [7] there are many more lines and there is more overlap.*

Energies of the strong Kr Auger lines [8]

$^2D_{5/2}$: 24.20, 29.60, 31.11, 37.84, 40.63, 51.33, 53.61, (55.43) eV
$^2D_{3/2}$: 25.45, 30.85, 32.35, 39.08, 42.31, 52.58, (56.69) eV

The figures in parentheses are for lines unlikely to be fully resolved.

Resonance energies for monochromator calibration

To calibrate X-ray monochromators the most useful features are the sharp and intense pre-edge resonances in absorption spectra of atoms or simple molecules with π^* or similar normally unoccupied orbitals. These resonances can be located by following the electron or ion yield as the photon energy is scanned. The energies given below, which are widely used for calibration, are compiled from multiple sources.

Energies of the first sharp pre-edge resonances in atoms and molecules

	Resonance (eV)	Excitation		Resonance (eV)	Excitation
Xe	65.110	$4d^{-1}6p$	N_2	400.865	$N1s^{-1}\pi^*(v=0)$
Kr	91.200	$3d^{-1}5p$	CO	534.2	$O1s^{-1}\pi^*$ (v unresolved)
Ar	244.39	$2p^{-1}4s$	Ne	867.12	$1s^{-1}3p$
CO	287.40	$C1s^{-1}\pi^*(v=0)$			

References

[1] A. Kramida, Yu. Ralchenko, J. Reader and NIST ASD Team (2016), http://physics.nist.gov/asd
[2] V. Schmidt, *Electron Spectrometry of Atoms using Synchrotron Radiation*, Cambridge University Press, 1997.
[3] M. Yu. Kuchiev and S.A. Scheinerman, *Sov. Phys. JETP*, **63** (1986) 986.
[4] O. Edqvist et al., *Physica Scripta*, **1** (1970) 25.
[5] P. Baltzer et al., *Phys. Rev. A*, **45** (1992) 4374.
[6] T.X. Carroll et al., *J. Electron Spectrosc. Related Phenom.*, **125** (2002) 127.
[7] L.O. Werme, T. Bergmark and K. Siegbahn, *Physica Scripta*, **6** (1972) 141.
[8] H. Aksela, S. Aksela and H. Pulkkinen, *Phys. Rev. A*, **30** (1984) 2456.

Appendix 3
Theoretical calculation of ionisation energies and spectra

The purpose of the first part of this appendix is to offer experimentalists an idea of the meaning of the 'terms of art', notably acronyms, used by theoreticians engaged in calculations of molecular electronic structure. Ideally, we might be able to judge the probable accuracy of calculated ionisation energies and spectral intensities according to the theoretical methods employed, but in reality, we must usually accept the word of the theoreticians. The following sections may seem very elementary to molecular physicists, but we hope they may be helpful or of interest to researchers who are less familiar with quantum-mechanical techniques.

There are three major classes of theoretical method: the Hartree–Fock (HF) or self-consistent field (SCF) method and its many extensions to include (more) electron correlation; density functional theory (DFT); and Green's function (GF) techniques. Calculations by all methods start from a basis of orbitals, that is, *one-electron* wavefunctions. The molecular orbitals themselves are constructed from *basis sets*, related to, but not identical with atomic orbitals. The functions in a basis set are usually centred at the positions of atoms in a molecule and mimic the angular and radial form of real atomic orbitals using mathematically convenient (Gaussian) functions; a linear combination of the atomic functions with adjustable coefficients is used to create molecular orbitals. All methods use the Born–Oppenheimer separation of nuclear and electron 'motions', and most use a non-relativistic Hamiltonian (Schrödinger equation), representing electron spins as an anti-symmetrised product with the spatial wavefunctions, creating spin orbitals. The simpler methods include only some of the real many-body electron–electron interactions which determine an important part of the energy, so additional strategies must be used to include as much of the missing interactions ('correlation') as possible. These basic ideas and much more are expounded in standard texts and particularly clearly by Szabo and Ostlund [1]. Almost all authors emphasise that the proper inclusion of electron correlation is even more vital in calculations on doubly ionised molecules than on neutrals, and this is the area where there are major differences between the available methods.

To calculate ionisation *spectra* rather than just transition energies, additional theory is ideally needed, involving the specific interaction of molecules with photons or other particles and coupling to the continua, but in practice this is almost never done. It is particularly difficult to model the two-electron continua involved in double photoionisation.

HF, SCF and ΔE-SCF

The self-consistent field (SCF), also known as the Hartree–Fock (HF) method, is based on the assumption that the many-electron wavefunction of the ground state of a molecule can be expressed as a single Slater determinant built from one-electron spin orbitals. Applying the variational principle, the complicated many-electron problem is reduced to a set of one-electron equations in which the electron–electron repulsion is treated in an average way, introducing the so-called direct and exchange Coulomb potentials.

Starting with a set of trial spin orbitals and varying the linear coefficients of their atomic constituents, the set can be solved iteratively to determine the optimal spin orbitals which minimise the total energy of the system. When no further improvement is possible by either expanding the basis set or iterating further, the HF limit has been reached. This is the point, indeed the only point in the whole story, where the orbitals are well defined and build up a single electron configuration for each electronic state. To calculate the electronic energies of states in a photoelectron spectrum it is possible to use Koopmans' theorem, which states that if all the remaining spin orbitals remain unchanged when one electron is removed (frozen orbital approximation), the energy needed to remove that one electron from a chosen orbital is equal to its HF binding energy (the eigenvalue of the SCF Hamiltonian) with just a sign change from negative to positive. Unfortunately this is a poor approximation, though still a very useful one, because it includes neither the correlation energies nor the relaxation energy, a consequence of the effect of loss of one electron on all the other electrons. When one electron is removed, all the orbitals change their form in response. The relaxation energy can be allowed for by recalculating the energy of each ionised state, again ideally to the HF limit in what is sometimes called a ΔE-SCF or Δ-SCF procedure. This is better, but still imperfect, as the correlation energy and any changes in it consequent upon each ionisation are neglected. If two electrons are removed instead of just one, the problems become more severe.

When the neutral ground state and all the ionised states are calculated separately, the potential energy surface of each one can also be calculated by repeating the whole set of calculations for different nuclear positions. The vibration frequencies of all the states and the shapes of the bands in a photoelectron spectrum can then be calculated from the vibrational overlap integrals (Franck–Condon factors) between states. Because this is such a demanding procedure it is seldom carried out, but in principle it can be done just as well for double as for single ionisation. Even then, it does not predict the relative intensities of electronic state bands, as no representation of the light field or of the electron continua has been introduced.

Beyond Hartree–Fock

In the HF model the interactions between electrons are calculated in terms of an effective average potential experienced by an electron due to the presence of all the other electrons. This is equivalent to using the time-averaged spatial distributions, but without inclusion of terms in the wavefunctions such as r_{12} expressing the instantaneous relative positions of a pair of electrons. The difference between the true (strictly exact non-relativistic) energy and the energy calculated at the HF limit is called the *correlation energy*; its magnitude is very significant, and to get accurate ionisation energies it must be estimated for both the ground state and the final state of each transition. The conceptually simplest method to do this is called configuration interaction (CI), and consists of expanding the wavefunction for each state as a linear combination of wavefunctions for different electron configurations of the same symmetry. The ground state, for instance, will be represented by the canonical HF orbital configuration plus many configurations with different numbers of electrons promoted to normally unoccupied orbitals. It can be shown that if an exponentially large number of configurations is included, the whole correlation energy is captured correctly. In practice such enormous expansions are impractical, so the theoretician must somehow choose the most efficient combinations. The choice of configurations gives rise to more acronyms; CASSCF (complete active space SCF) means including all possible configurations involving a set of orbitals defined as the active space, which could be, for instance, all valence orbitals and their virtual counterparts. It turns out that the canonical HF orbitals are not the most effective choice as the basis for a finite CI expansion, so in MCSCF (multi-configuration SCF) methods both the orbitals and the expansion coefficients in the truncated CI expansion are optimised together; this is the method used in many of the 'high-level calculations' referred to in earlier chapters. CASSCF calculations are often improved further by additional configuration interaction in a technique called CASSCF-MRCI (CASSCF-multi-reference CI), which has been called the minimum requirement for satisfactory representation of double ionisation spectra [2]. Such calculations achieve an absolute accuracy of a few tenths of an eV for energies of the main two-hole (2h) low-energy doubly ionised states of small to medium-sized molecules.

A different method of improving upon the HF model is to use perturbation theory, usually known in this context as Møller–Plesset theory. It starts from the known eigenfunctions and eigenvalues of the HF Hamiltonian and includes correlation corrections as an infinite series of matrix elements between the original eigenfunctions and a perturbation operator, the electron–electron interaction (${r_{12}}^{-1}$ in atomic units). This is an efficient computational method, which has been used to calculate several double ionisation energies [3] and has more recently been found useful as an adjunct to MCSCF and other methods in calculating the energies of core–valence doubly ionised states [4].

Another form of theory going beyond HF and capable of highly accurate predictions of double ionisation energies is called coupled-cluster (CC) theory. The effect of excitations from the basic configuration is represented by excitation operators, which may include single, double, triple or higher excitations with corresponding acronyms CCSD, CCSDT and so on. Triple excitations are the most computationally demanding, and they may be included by a perturbation method instead; the acronym for this method is CCSD(T). There is also a multi-reference form (MRCC). There are several variants of this method, many of them described in an article by Musial et al. [5], where CC theory has been used to calculate double ionisation energies for ground and low excited states of a number of small molecules.

An important strength of these HF-based methods is that as with HF alone, the whole potential energy surfaces of ground and ionised states can be calculated accurately if sufficient computing time and power is available. This is important for studies of the dissociations and other reactions of ions, but is also relevant to spectra, since without such extensive calculations band shapes can only be estimated.

Density functional theory (DFT)

The basic idea of DFT, like that of the earlier Thomas–Fermi model, is to replace the external potential in the Schrödinger equation by the electron density. The true electron density as a function of position, with its cusps and gradients, defines the positions and charges of the nuclei, so knowledge of it is sufficient to completely determine the electronic structure of the ground state of a molecule. In particular, the ground-state energy can be expressed as a functional of this density which depends on just three spatial coordinates instead of the many coordinates for the positions of all electrons, and the correct electron density can be found through minimisation of the energy. The substitution of the electron density for the complete many-electron wavefunction leads to an immense reduction of computational demands, making the calculations manageable even for very large molecules where other theoretical methods such as CI are impractical. If the true expression for the energy functional which uses the electron density were known, the electronic structure problem for a many-body system would be completely solved. In reality the true functional is not known so approximate functionals, optimised for different purposes, are used. After several theoretical advances, the method has become very widely used and effective for calculating the energies of ground states. It is seldom used for excited states, so is effective for calculation of the lowest double ionisation energies of large molecules, but is less useful for complete spectra. The accuracies of different forms of the theory (different functionals) in calculating double ionisation energies have been evaluated for a good number of atoms and molecules [6].

Green's function (GF) methods

Green's function methods are the most powerful tools currently available for the calculation of double ionisation spectra and have been applied to many of the molecules treated in preceding chapters. For calculations on double ionisation, the two-particle many-body Green's function must be evaluated. The calculations start from the HF orbital description of a molecular ground state and the double ionisation energies are obtained as solutions of a Hermitian eigenvalue problem as energies of the poles. The corresponding eigenvectors contain information about the pole strengths, which are related to spectral intensities. The use of the HF one-electron basis makes it possible to analyse the eigenvectors based on their expansion in terms of two-hole (2h) and three-hole-one-particle (3h1p) configurations relative to the ground state and, therefore, to assign them to specific electronic configurations. For mathematical details, the reader would need to turn to the original work of Cederbaum and his group [7–9], who developed the method as it is applied to molecular ionisation. It was already the leading method for single ionisation spectra from the 1970s, and was developed particularly for double ionisation somewhat later. To create spectra for double photoionisation, the spectrum of calculated pole strengths is often folded with a Gaussian representing a combination of experimental resolution and the unknown Franck–Condon envelopes of the states. To represent Auger double ionisation spectra, where singlet states are populated preferentially, it is usual to triple the pole strength of 2h singlet states.

Relativistic calculations

For some molecules containing heavy atoms such as iodine or mercury, none of the non-relativistic methods are fully satisfactory. Corrections for relativistic effects such as spin-orbit interaction can be applied to non-relativistic calculations and are sufficient in cases where the effects are minor or localised, but where they are strong a different form of calculation should be adopted. Instead of the familiar orbitals, the basic functions are four-component *spinors* and the usual Hamiltonian is replaced by the Dirac–Coulomb or another relativistic operator. The mysteries of these advanced calculations can safely be left to the theoreticians, but as users we should be aware that description of the resulting spectra of states in terms of orbital configurations is no longer appropriate, and the familiar term symbols (Mulliken symbols) also must be jettisoned. All types of calculation, including both MCSCF and Green's function calculations can be done in fully relativistic forms. Different relativistic methods applied to double photoionisation spectra are presented in papers on Br_2^{2+} [10] and CH_3I^{2+} [11].

Thermochemical calculations

To go from relativistic quantum theory to thermochemistry may seem like descent from the sublime to the ridiculous, but the energies of dissociated states of doubly charged ions are calculated more accurately by thermodynamics, when the relevant quantities are known, than by any extant quantum theory. Thermodynamic calculations are highly relevant where a doubly charged ion has no bound states, as with $[H_2O]^{2+}$, $[CF_4]^{2+}$, and similar species, especially if indirect double ionisation processes are possible. The fundamental equation is an expression of the First Law of Thermodynamics, and the energy ΔE_{ij} required to form separated ions $m_i^+ + m_j^+$ with no neutral fragments is just

$$\Delta E_{ij} = \Delta H_f\,(m_i^+) + \Delta H_f\,(m_j^+) - \Delta H_f\,(m_{ij})$$

where $\Delta H_f(m)$ is the heat of formation of the ion or neutral species in question. This works because all heats of formation are referenced to a common standard state of the constituent elements. An alternative formulation for the case where a single bond between two molecular fragments X and Y is broken is

$$\Delta E_{ij} = D\,(X - Y) + IE\,(X) + IE\,(Y)$$

where D(X−Y) is the dissociation energy of the bond between X and Y and IE(X) and IE(Y) are the ionisation energies of the separate fragments. More generally, and necessarily if there are many fragments

$$\Delta E = \Sigma \Delta H_f\,(\text{all products}) - \Delta H_f\,(\text{starting molecule}).$$

The heats of formation of both ions and neutrals, and dissociation energies, can be obtained from online sources such as NIST [12] or from a very useful compilation in book form [13]. The values used should be those for a temperature of 0 K, which exclude internal energy content if they are available. As an example, we calculate the minimum energy for formation of $C^+ + O^+$ from CO, which is closely approached in indirect ionisation of that molecule by dissociation followed by autoionisation of O* (Chapter 2, Section 2.4).

The 0 K heats of formation of CO, C^+ and O^+ are given in ref. [13] as –27.2, 429.7 and 373.0 kcal mol^{-1} or -110.5, 1803.2, and 1563.1 kJ mol^{-1}. The minimum energy for appearance of the ion pair is therefore 829.9 kcal mol^{-1} (3,472.1 kJ mol^{-1}) or 35.99 eV. Thus the minimum double ionisation energy at which $C^+ + O^+$ can possibly be formed from CO is 36 eV, if both the ejected electrons and the ions formed in their ground states have zero kinetic energy. In the case of CO, the most intense O* autoionisation gives electrons of about 0.4 eV energy, so the minimum single ionisation energy for this particular process is 36.4 eV. The actual onset is seen in Figs. 2.4.2 and 2.4.3 at a double ionisation energy just above 36 eV, possibly at 36.4 eV, implying that the ions have almost zero kinetic energy at the lowest onset. Since some kinetic energy will always come from Coulomb repulsion after O* autoionisation, then the ion pair must have been born in those events at a considerable distance apart, of the order of 100 Å. While CO is an extreme case, it is worth emphasising that where there is no bound state of a molecular doubly charged ion accessible by a vertical transition from the ground-state molecule, indirect double ionisation pathways may still give access to double ionisation at a lower energy than the true vertical molecular process. If bond lengths expand even a little towards dissociation in an intermediate singly ionised state, the effect on the minimum double ionisation energy is magnified by the steepness of the Coulomb repulsion curve (potential energy surface) of a dissociative double ionisation. When the determination of ionisation energy is based on electron pairs alone, such an effect is indistinguishable from an ionisation energy lower than the true vertical one. If ions are detected in coincidence with the electron pairs, as can be done in more complex experiments [14], it may be possible to measure their kinetic energies and use them to correct the thresholds determined from electron energies alone.

Empirical calculations

Valence double ionisation energies

We have used empirical calculations at many points in this book to help with assignments, and they have been even more useful at experimental stages in predicting the range in which to expect double ionisation energies. The empirical formulae are summarised here for convenient reference.

The first and simplest formula is the 'rule of thumb' [15] which relates the lowest double ionisation energy to the lowest single ionisation energy of the same molecule:

$$\mathrm{DIE} = 2.8\,\mathrm{IE}.$$

This works surprisingly well for a wide range of closed-shell atoms as well as molecules. The factor of 2.8 can be refined for particular classes of related molecules such as aromatic hydrocarbons [16]. The basic idea behind this formula is that when one electron is removed, the energy needed to remove another one is increased partly by the tighter binding of all electrons and partly by the Coulomb repulsion energy between the two final charges. For molecules which contain two or more functional groups on which the positive charges are likely to be located, and for rigid molecules where the final intercharge distance can be estimated, it is reasonable to separate these two factors. An expression of the form

$$\mathrm{DIE} = \mathrm{k}\,\mathrm{IE} + \mathrm{g}/r_{12}$$

with k and g as constants was fitted to a number of reliably known double ionisation energies. Using r_{12} values estimated in a straightforward way from the molecular dimensions, the formula

$$\mathrm{DIE} = 2.2\,\mathrm{IE} + 11.5/r_{12}$$

with energies in eV and distances in Å units was found to represent the double ionisation energies of some 50 molecules much more closely than the simpler formula [17]. Coulomb repulsion means that the charges will be as far apart as possible in the lowest energy state of a doubly charged ion; for a floppy molecule, this may mean that in the dication an open-chain conformer is favoured as against a more compact form in the neutral molecule. We note that this situation calls into question the basic definition of the molecular double ionisation energy.

Core–core and core–valence double ionisation energies

For core–core double ionisation in the 1s shells, the empirical hydrogenic model introduced in Chapter 6 to estimate the ionisation energies is

$$\mathrm{IE}(1\mathrm{s}^{-2}) = 2\,\mathrm{IE}(1\mathrm{s}^{-1}) + 11.7\mathrm{Z}$$

where Z is the atomic number of the atom being ionised. This model still needs testing on lighter and heavier atoms to evaluate its general validity.

For core–valence double ionisation, the scheme of interpreting core–valence (CV) spectra espoused in Chapter 6 is equivalent to a model where the CV ionisation energy CV(j) to the jth singlet-triplet pair is given by

$$CV\,(j) = E\,(C) + IE\,(j) + e^2/r_{12}$$

where E(C) is the energy at the edge for producing a single hole and IE(j) is the valence ionisation energy for removing an electron from the jth orbital in the neutral molecule. As all the other quantities are known or measured, this model allows us to estimate the intercharge distance r_{12} in each final state. The values that emerge for intercharge distances in the lowest CV ionised states are quite reasonable; for the atom-like molecules CH_4, NH_3 and H_2O, the apparent r_{12} values are all about 1 Å. For the 'spherical' molecules CF_4 and SF_6, the apparent distance between the charges is 2 Å, for CCl_4 and benzene 3 Å and for C_{60} 7 Å. This is such a crude model, however, that it seems inappropriate to push it further.

References

[1] A. Szabo and N.S. Ostlund, *Modern Quantum Chemistry*, McGraw-Hill, New York, 1989.
[2] F.R. Bennet, *J. Chem. Phys.*, **103** (1995) 3613.
[3] W.J. Griffiths, *J. Mass Spectrom. Ion Processes*, **112** (1992) 45.
[4] J. Niskanen et al., *Phys. Rev. A*, 82 (2010) 043436.
[5] M. Musial, A. Perera and J. Bartlett, *J. Chem. Phys.*, **134** (2011) 114108.
[6] D.P. Chong, *J. Chem. Phys.*, **128** (2008) 084112.
[7] L. Cederbaum et al., *Adv. Chem. Phys.*, **65** (1986) 115.
[8] J. Schirmer and A. Barth, Z. Phys. A, **317** (1984) 267.
[9] F. Tarantelli, *Chem. Phys.*, **329** (2006) 11.
[10] T. Fleig et al., *Chem. Phys.*, **343** (2008) 270.
[11] M. Pernpointner, *Chem. Phys.*, **407** (2012) 39.
[12] http://webbook.nist.gov
[13] S.G. Lias et al., Gas Phase Ion and Neutral Thermochemistry, *J. Phys. Chem. Ref. Data*, **17** (1988), Supplement 1. Published by the ACS, AIP and NBS.
[14] J.H.D. Eland and R. Feifel, *Chem. Phys.*, **327** (2006) 85.
[15] B.P. Tsai and J.H.D. Eland, *Int. J. Mass Spectrom. Ion Phys.*, **36** (1980) 145.
[16] S. Tobita et al., *Can. J. Phys.*, 72 (1994) 1060.
[17] R.D. Molloy et al., *Chem. Phys.*, **335** (2007) 4267.

Index

Printed and bound by CPI Group (UK) Ltd, Croydon, CR0 4YY